AF600257

BORDERLINE CANADIANNESS

Border Crossings and Everyday Nationalism in Niagara

JANE HELLEINER

Borderline Canadianness

Border Crossings and Everyday Nationalism in Niagara

UNIVERSITY OF TORONTO PRESS
Toronto Buffalo London

Toronto Buffalo London
www.utppublishing.com

ISBN 978-1-4426-4905-7 (bound)

Library and Archives Canada Cataloguing in Publication

Helleiner, Jane Leslie, 1961–, author
Borderline Canadianness : border crossings and everyday nationalism in Niagara / Jane Helleiner.

Includes bibliographical references and index.
ISBN 978-1-4426-4905-7 (bound)

1. Nationalism – Ontario – Niagara Peninsula. 2. Globalization – Ontario – Niagara Peninsula. 3. Boundaries – Social aspects – Canada. 4. Ethnology – Ontario – Niagara Peninsula. 5. Niagara Peninsula (Ont.) – Social conditions. 6. Canada – Boundaries – United States. 7. United States – Boundaries – Canada. 8. Canada – Relations – United States. 9. United States – Relations – Canada. I. Title.

FC3099.N53Z62 2016 971.3′38 C2016-901216-6

University of Toronto Press acknowledges the financial assistance to its publishing program of the Canada Council for the Arts and the Ontario Arts Council, an agency of the Government of Ontario.

Canada Council for the Arts Conseil des Arts du Canada

Funded by the Government of Canada Financé par le gouvernement du Canada

Canada

Contents

Acknowledgments

This book would not have happened without the assistance of many people. My primary debt is to the Canadian Niagara residents who responded to the invitation to share their experiences of border life. My colleagues Joan Nicks and Jeanette Sloniowski offered early enthusiasm and insights, and my ability to keep the project moving forward depended on research assistants Carolina Alvarado, Allison Burgess, Jessica Craig, Lindsey Foley, Elizaveta Kozlova, Cynthia Nyarko, Betty Scott, Katie Sutton, Melissa St. Germaine-Small, and Karen Wiens. Cynthia Nyarko's MA thesis, "Canada/U.S. Border Crossing: Facilitation and Constraint" (2004), is based on some of the interviews.

My colleagues in the Department of Sociology as well as the MA in Social Justice and Equity Studies at Brock University have ensured a collegial environment for many years. Particular thanks are due to June Corman, Kate Bezanson, David Butz, Mary Beth Raddon, and Susan Tilley, whose administrative leadership allowed me to focus on research and writing. Janet Conway provided a Toronto retreat, and Nancy Cook, Margot Francis, Amanda Glasbeek, Hijin Park, Anna Pratt, Rebecca Raby, and Michelle Webber supplied various forms of advice and feedback on this project.

The decision to turn the research into a book would not have happened without the encouragement of Douglas Hildebrand at the University of Toronto Press. The constructive input of three anonymous reviewers for the Press greatly strengthened the manuscript. Initial research funding came from a Social Sciences and Humanities Research Council Institutional Grant from Brock University (2000), and then a Social Sciences and Humanities Research Council of Canada Standard Research Grant #410-2001-0894 (2001–4). Book publication was made possible by funds from the Council for Research in Social Sciences (2013) and BSIG/BUAF Special Purpose Grant (2015) and the Department of Sociology at Brock University. I am grateful to Associate Dean of Social Sciences Diane Dupont for her assistance.

I acknowledge below permission to use portions of previously published material. The first part of chapter 2 includes material that appeared earlier in Helleiner (2007), "'Over the River': Border Childhoods and Border Crossings at Niagara," *Childhood: A Global Journal of Child Research* 14 (4): 431–47 (doi: 10.1177/0907568207081850). Parts of chapter 3 previously appeared in Helleiner (2010), "Canadian Border Resident Experience of the 'Smartening' Border at Niagara," *Journal of Borderlands Studies* 25 (3&4): 87–103. Chapter 4 includes reworked material from the above as well as Helleiner (2012), "Whiteness and Narratives of a Racialized Canada/US Border at Niagara," *Canadian Journal of Sociology* 37 (2): 109–35. Portions of chapter 5 are derived, in part, from Helleiner (2009), "Young Borderlanders, Tourism Work, and Anti-Americanism," *Identities: Global Studies in Culture and Power* 16 (4): 438–62 (doi: 10.1080/10702890903020950), as well as from Helleiner (2009), "'As Much American as a Canadian Can Be': Cross-Border Experience and Regional Identity among Young Borderlanders in Canadian Niagara," *Anthropologica* 51 (1): 225–38. Some of chapter 6 is also derived in part from Helleiner (2013), "Unauthorised Crossings, Danger and Death at the Canada-US Border," *Journal of Ethnic and Migration Studies* 39 (9): 1507–24 (doi: 10.1080/1369183X.2013.815431).

Long-time reading and writing friends Linda Carson, Susan Hughes, and Klari Kalkman have patiently listened to me talk about this work for many years. My parents, Gerry and Georgia; brother, Eric; sister-in-law, Jennifer; and niece and nephew, Zoe and Nels, have helped me think about border crossing, while my brother, Peter, has ensured that I have taken regular breaks from writing – in beautiful surroundings. My sons, Kieran and Tomas, provided research assistance and, along with my daughter-in-law, Katharina, and grandchild, Jannik, have filled my life with joy. Finally the unwavering intellectual and loving support of my partner, Bohdan, ensured both the launch and the successful completion of this project. Thank you all.

BORDERLINE CANADIANNESS

Border Crossings and Everyday Nationalism in Niagara

Introduction

Studies of contemporary Canadian nationalism often emphasize how this is forged in juxtaposition to a more powerful U.S. neighbour, and the Canada/U.S. border is occasionally portrayed as an exemplar of the asymmetrical relationship. During the Canadian 1988 election, fought in large part over what would become the 1989 Canada-U.S. Free Trade Agreement, the then opposition Liberal Party, for example, aired its famous "erasing the border" political ad. This ad portrayed a U.S. official engaged in free trade negotiations with his Canadian counterpart stating, "There's one line I'd like to change." As the Canadian asked, "Which line is that?" an eraser was shown rubbing out the Canada/United States border on a drawn map of North America. As the erasing continued, the U.S. negotiator replied, "This line here. It's just getting in the way." The portrayal of free trade as threatening Canadian sovereignty through border erasure was rhetorically effective, but the re-election of the pro–free trade Conservatives led to the adoption of the Free Trade Agreement in 1989 and then the trilateral (Canada, U.S., and Mexico) North American Free Trade Agreement (NAFTA) in 1994.

The extent to which these free trade agreements actually precipitated processes of Canada/U.S. "de-bordering" was taken up by scholars from a variety of perspectives, but the expanding research on the "thinning" of this border had to be revisited when the events of 9/11 led to an increased securitization of the Canada/U.S. borderline. As Canadian media featured long lines of U.S.-bound commercial traffic stalled at major land border crossings, scholars began to talk of Canada/U.S. "*re*-bordering," and some Canadian leaders voiced concern that U.S. border closure rather than erasure was the more significant threat to Canada.

In this book I approach a changing Canada/U.S. border from the point of view of those living in Canadian Niagara, one of Canada's most significant border regions. Drawing upon interviews with border inhabitants as well as local press reporting, I document and analyse life at the Niagara Canada/U.S. borderline with the goal of contributing to scholarship on borders, bordering, and nationalism in a globalized world.

My interest in border life in Canadian Niagara began with my move to the region from Toronto in the early 1990s. Upon arrival, I was immediately struck by the salience of the local Canada/U.S. border to everyday life on the Canadian side, as new neighbours and colleagues welcomed me to the area by offering information about going "over the river" to the U.S. for shopping and leisure activities. The prevalence of "border talk," in my early interactions with Canadian border residents reminded me of childhood visits to my maternal grandparents, who lived in the Southwest Ontario border city of Sarnia. Memories of my grandparents' home facing the St. Clair River with a direct view across to Port Huron, Michigan, came back to me, as did my mother's stories of growing up with cross-border shopping and early employment serving U.S. visitors. As I began to settle into my new home, I became increasingly interested in how Canadian Niagara residents experienced and understood border life.

Running the length of 8,891 kilometres (including the Canada/ Alaska sections), the Canada/U.S. border is a significant one by global measures, with nearly 300,000 people and $2 billion in goods and services crossing daily ("Canada-US Trade Relations: What's Next for the Pivotal Partnership?" *Globe and Mail*, 26 April 2015, B3). Approximately 80 per cent of the Canadian population lives within 160 kilometres of this borderline – a fact that has led some scholars to describe Canada in its entirety as a "borderland society" (Gibbins 1989). Canadian Niagara, however, as indicated above, seemed to me to be much more of a salient "borderland" than Toronto.

Within the diversity of Canada/U.S. border regions, Niagara is unique as a global tourist destination with its spectacular Niagara Falls and surrounding landscape that have inspired voluminous artistic, literary, travel, journalistic and cinematic representation.[1] Niagara is also one of the most densely populated and intensely traversed of Canada/U.S. border regions, with four vehicle and three railway (two currently functional) bridges traversing the Niagara River, which marks the international boundary line. The vehicle bridges consist of the Lewiston-Queenston (or, for Canadians, Queenston-Lewiston) Bridge that links the two villages of Queenston, Ontario, and Lewiston, New York; the Whirlpool Rapids and Rainbow Bridges that both connect the cities of Niagara Falls, Ontario, and Niagara Falls, New York; and the Peace Bridge that joins the town of Fort Erie, Ontario, to the major urban centre of Buffalo, New York.

The Peace Bridge is the second busiest Canada/U.S. border crossing after the Ambassador Bridge linking Windsor and Detroit. Although the Ambassador Bridge has a higher volume of truck traffic, the combined Niagara bridges have the highest numbers of vehicle, bus, and pedestrian crossings. In 2010, over 13 million people crossed Niagara bridges, and the value of imports and export exceeded $60 billion (Binational Economic and Tourism Alliance 2011). As a major North American transportation and trade corridor as well as a global tourist destination, Niagara is often featured in official Canadian border-related discourse and policy.

Thoughts of conducting research into border life in Canadian Niagara percolated in my mind for several years prior to my beginning this study. The decision to finally formulate a research project, however, was precipitated by a national press report about a tragic case at the Rainbow Bridge. The report described how a U.S. customs officer at the Rainbow Bridge had pulled over a vehicle for inspection, after noticing an empty infant car seat, and found a baby and his mother under a blanket on the floor of the back seat. The child (who was identified as South African in a subsequent news story), was later pronounced dead, and the mother and the two other adults in the vehicle (the latter identified as legal immigrants to the U.S. from South Africa) were charged ("Three Charged in Baby's Death," *Globe and Mail*, 21 March 1998, A9; "Desperate Crossing," *Globe and Mail*, 26 February 1999, A3). These press stories of an apparent case of attempted unauthorized migration resulting in an infant death made a strong impression on me, perhaps because of resonances with another part of my family biography, notably my own father's harrowing journey as a child migrant fleeing Nazi occupation.

My new realization that the Niagara border was a site of global migration as well as local crossings, continental commercial flows, and tourism, made me think more deeply about how Canadian border residents were engaged with a multiscalar border in a world where, as Paasi has observed, it is "increasingly difficult to think of certain borders as local and others as global" (as quoted in Johnson et al. 2011, 62–3).

A renewed interest in Canadian Niagara resident experiences of border life led to the decision to begin this study in the spring of 2001. I launched the project by conducting two pilot interviews with two older adult colleagues who were long-time border residents. I had conducted four additional interviews with younger adults and had others lined up when the events of 9/11 occurred. As U.S. and Canadian government responses to these events directly impacted the Niagara border, it became clear that my newly launched project would be shaped by the new context, and I worked to widen its scope to include post-9/11 experiences. I continued with the research and by 2004 had conducted fifty-one interviews with Canadian Niagara residents and amassed over one thousand local, border-related newspaper reports.

Borders and Bordering

This study draws from and contributes to a flourishing interdisciplinary and comparative border studies that aims to "to chronicle and understand how borders, border cultures, societies, polities and economies, are not only changing … but also how borders often play key roles in these changes" (Wilson and Donnan 2012a, 11).[2] This field of study has emphasized how territorial borders mark and enclose the state and nation while simultaneously facilitating and regulating binational and more globalized flows. In the 1980s and 1990s much of the border studies scholarship was animated by the apparent "de-bordering" that accompanied the end of the Cold War, the collapse of the Soviet Union, the emergence of the European Union, and the North American Free Trade Agreements already mentioned. The heightened U.S.-led border securitization that followed the events of 9/11, however, prompted a subsequent wave of research documenting and analysing the dynamics of what appeared to be a new phase of global "re-bordering" (Andreas and Biersteker 2003).

This study draws upon an ethnographic tradition within border studies that has provided grounded "bottom-up" descriptions and analyses of border life and is particularly inspired by work focused on border residents' "local narratives" of everyday border life (Paasi 2011, 6). According to Wilson and Donnan (2012a) ethnographic research that prioritizes "how borders are constructed, negotiated and viewed from 'below" (8) continues to provide "much of the lifeblood of border studies" (13).[3]

Approaching the study of borders through the local narratives of border residents offers an important intervention, given the tendency of much border-related research to reflect the priorities of central states and other powerful political and economic players. Considerable European-based border research, for example, is funded by the European Union, and in the 1990s, some support for U.S. border research came from trade organizations and NAFTA (Wilson and Donnan 2012a; Newman 2011, 40). Since 9/11, U.S. Department of Homeland Security funding associated with an expansive "homeland security–industrial complex" has supported networks of "corporate and bureaucratic entrepreneurs" and their political partners, while researchers at "universities and think tanks ... lurk underfoot in hopes of a few research crumbs" (Heyman 2008, 320).

Border research funded by dominant interests can lead to studies of border regions and peoples that reflect the preoccupations of the powerful. Being more autonomous of such funding considerations, I was fortunate in that I could design a project focused on learning about border life from the perspective of border residents. The resulting "bottom-up" study, inspired by the ethnographic tradition mentioned above, usefully augments existing scholarly work on the contemporary Canada/U.S. border, much of which comes out of what Paasi (2011,16) has described as more macro-level political economy and geopolitical approaches within border studies. Post-9/11 scholarship on the Canada/U.S. land border in particular is largely focused (both supportively and more critically) on the border discourses and projects of national as well as regional economic and political elites (Brunet-Jailly 2007, 2012; Gilbert 2007, 2012; Nicol 2011; Salter 2007; Salter and Piche 2011). This case study, by contrast, focuses on the everyday experiences of those for whom the border region is home rather than a corridor for, or site of, continental transportation, trade, tourism, or securitization.

While much of the research in border studies, like this study, addresses state territorial borders, newer work on mobilities and migration has broadened the field by documenting how territorial boundaries operate in conjunction with bordering processes at a much wider range of sites and scales. As this scholarship reveals, for example, states may engage in forms of "remote control" – by regulating who can and cannot easily and/or legitimately approach territorial borders (e.g., through visa requirements or marine interdiction) – as well as forms of internal bordering (e.g., through immigration and workplace enforcement against non-citizens).[4] While my study is centred on a particular territorial border, this broader approach to bordering allows me to situate the territorial Niagara border within wider interconnected multiscalar bordering processes that shape this border as well as the people and goods that flow across it.

The newer work on bordering is also useful to this study, because it highlights some of the limits of the de-bordering/re-bordering paradigm. In particular, newer work reveals that territorial and non-territorial bordering processes are more complex than this paradigm suggests. Rather than simply opening up or closing down flows, bordering processes produce a filtering of flows as, for example, the cross-border mobilities of people and/or goods are eased or constrained, sped up or slowed down, in deeply differentiated ways. Borders in this context, then, function more as sieves than walls or gates.

Some of the research on filtering at territorial and non-territorial borders goes beyond documenting such "sieving" processes to ask further about the effects of differentiated and unequal cross-border im/mobilities in a globalized world. Some of the relevant scholarship for example, connects such unequal im/mobilities to the reproduction of social inequality within and across regional, national, and global spaces (e.g., Lugo 2000; De Genova 2002; Cunningham and Heyman 2004; Van Houtum 2010). In this book, I bring these insights to my analysis of local narratives, as I consider how the accounts of border life gathered from border residents living in Canadian Niagara illuminate processes of filtered bordering at the Canada/U.S. border as well as their local and wider impacts.

In focusing on filtered bordering and its effects, this study draws on a comparative border studies, but it also builds on and contributes to existing research on Canadian territorial and non-territorial bordering that challenges still dominant constructions of a benign and even "friendly" Canadian border (e.g., Grinde 2002; Sharma 2006a, 2006b; Bhandar 2008; Pratt and Thompson 2008; Razack 2010; Mountz 2010; Simpson 2014). The examination of life in Canadian Niagara offers new insights into a changing Canada/U.S. border, filtered bordering processes, as well as constructions of borderline Canadianness.

Borderline Canadianness

As indicated at the outset, scholarship on Canadianness has long emphasized how Canadian identity is forged in juxtaposition to a hegemonic Americanness (e.g., Mackey 1999; Kymlicka 2003; Winter 2007). The relationship between Canadianness and the Canada/U.S. border itself has been explored in a range of literary, cinematic, photographic, and journalistic representations. Some of these representations highlight the impact of the borderline on border residents.[5] There is, for example, recurrent Canadian media reporting on the towns of Stanstead, Quebec, and Derby Line, Vermont, where the Canada/U.S. boundary line famously runs through a library, an opera house, and some private residences. Other commonly depicted communities include the U.S. exclave of Point Roberts, accessible only from Canada, and the Mohawk Nation at Ahkwesáhsne, upon which is imposed the international boundary line as well as the provincial and state jurisdictions of Ontario, Quebec, and New York State. On the one hand, portrayals of these more unusual border settings draw attention to the artificiality of state boundary making, but on the other hand, they reinforce dominant constructions of a self-evident and naturalized border by emphasizing their allegedly anomalous character.

Canadian cultural representations of the Canada/U.S. border point to the importance of this border for reflections on and constructions of nationalized Canadianness, but despite the characterization of Canada as a "borderland society," there remains limited scholarship on Canadian border life outside of the communities mentioned above. In this book, both interviews with border residents and local press reporting are used to expand understanding of the complexities of Canadianness as manifested in the everyday life of Canadian Niagara residents.

My exploration of how everyday Canadianness is experienced and imagined at the Canada/U.S. border parallels similar work on border nationalisms within comparative border research. The pathbreaking theorizing of Anzaldua (1999) and Rosaldo (1993) challenged scholars to understand "border areas as places of interpenetrating spaces and more complex, nonunitary identities" (Kearney 1995, 557), and subsequent work has clearly demonstrated how varied nationalized border identities further articulate in complex ways with region, class, race, and gender (e.g., Vila 2000, 2003, 2005). Following this work, I examine "everyday nationalism," understood as "the actual practices through which ordinary people engage and enact (and ignore and deflect) nationhood and nationalism in the varied contexts of their everyday lives" (Fox and Miller-Idriss 2008, 537) in Canadian Niagara.

The documentation and analysis of borderline nationalism is informed by Canadian studies scholarship that has addressed the varied ways in which Canadianness is constructed vis-à-vis Americanness (as discussed earlier) as well as a wider globalized world (e.g., Jiwani

2009; Jefferess 2009; Shahzad 2012). My approach also draws upon research that highlights the ways in which forms of Canadian nationalism may reflect and reproduce structures of white settler colonialism, as well as ethnoracialized, gendered, and classed inequality (e.g., Mackey 1999; Grinde 2002; Razack 2002, 2004; Thobani 2007; Henry and Tator 2010; Dhamoon and Abu-Laban 2009; Razack, Smith, and Thobani 2010; Simpson 2014). The case study of borderline Canadianness in Niagara, then, builds on and augments a rich body of work on Canadian nationalism.

Research Themes and Outline of Book

Inspired by and aiming to contribute to both border studies and Canadian studies, this book then focuses on three major themes. The first is how Canadian border residents experience and imagine their local border and some of the "top-down" central state and regional elite border projects enacted there. The interviews and local border press reports are used to illuminate the ways in which the border and border-related projects are variously supported, contested, and even subverted by border residents, thereby revealing how this border is constructed and engaged "from below" (Doevenspeck 2011).

The second theme reflects newer and more critical border studies and Canadian work on filtered bordering that highlights the significance of differentiated and unequal border im/mobilities. Following Heyman (2012), who has linked filtered bordering more specifically to "complex patterns of interrelated inequality in ... borderlands" (60), this study pays particular attention to how filtering at the border is linked to cohort and life course as well as to classed, gendered, Indigenous, ethnoracialized, and citizenship positionings. The significance of such filtered bordering for the production and reproduction of inequality in Canadian Niagara is a major focus of the study as the effects, it is argued, challenge constructions of homogenized "border culture" found in some official and scholarly descriptions of Canada/U.S. border communities (Turbeville and Bradbury 2005; Konrad and Nicol 2008, 2011; Konrad 2012).

The third theme draws again on border studies and Canadian studies in its focus on everyday nationalism and, specifically, the ways in which Canadian border residents enact and imagine their relationship to Canadianness. The study examines how Canadian border residents combine everyday *bi*nationalism – in the form of frequent border crossing – with varied and complex forms of nationalized belonging. The ways in which border resident constructions of Canadianness are further marked by lines of inclusion and exclusion are emphasized.

The three themes of local engagement with the border and top-down border projects, filtered bordering, and borderline Canadianness are woven through the book, which is organized as follows. In chapter 1, I contextualize the study through an introductory discussion of the history and political economy of Canada/U.S. bordering, before turning to a more localized account of the geography, history, and political economy of Canadian Niagara. This is followed by a review of the methodology of the study, including details of the interviews and local press reporting that provide the basis for the analysis.

Chapter 2 begins to explore Canadian border resident experiences and understandings of the local Canada/U.S. border through an examination of interviewee narratives of crossing "over the river" as children and youth in the pre-9/11 era. The focus on retrospective accounts of cross-border activities and border inspections provides insight into how young people growing up in a major Canadian border region experienced and understood state border projects and bordering processes. The documentation and analysis of growing up in the border region in the pre-9/11 period provide important context for accounts of post-9/11 border life, bordering, and Canadianness.

Chapter 3 combines interviewee accounts and press reporting to examine both official regional elite and more everyday responses to 9/11 and post-9/11 border securitization in Canadian Niagara. The material illuminates Canadian border residents' experiences of a changed border space and more surveilled cross-border mobilities. Border residents' critiques as well as legitimation of local border securitization are highlighted and the complexity of Canadian border residents' engagement with post-9/11 changes is emphasized.

Chapter 4 looks more carefully at interviewee accounts of filtered bordering, notably how classed, gendered, Indigenous, ethnoracialized, and citizenship positionings were linked to differentiated and unequal border crossing experiences. The analysis considers the significance of unequal im/mobilities in the context of everyday border life and the implications for official constructions of "border culture" in Canadian Niagara.

Chapter 5 pursues the discussion of border identity by drawing upon press reporting and the interview material to further explore the phenomena of official cross-border regionalism and everyday nationalism at the borderline. The intersections of nationalized identities with classed, ethnoracialized, and gendered positionings are considered, as is the phenomenon of anti-Americanism in the context of youth tourist employment and the experiences of Canadian/U.S. dual citizens.

Chapter 6 extends the discussion of border life, bordering and nationalism to consider borderline Canadianness in the context of globalization. This involves an analysis of interviewee and press constructions of global mobilities associated with tourism, multiculturalism, and migration with particular attention given to local constructions of unauthorized global migrant border crossings. Analysis of the ways in which national and global bordering are combined in border resident lives and imaginations is followed by an exploration of locally based counter-hegemonic border-related visions and activism. A concluding chapter provides a brief update on post-research Canada/U.S. and Niagara border developments and final thoughts on the contributions of the study.

Chapter One

Bordering Canada at Niagara

Introduction

In this chapter, I contextualize the Canada/U.S. border at Niagara within a broader discussion of the history and political economy of Canada/U.S. bordering before moving on to focus more specifically on the border region of Canadian Niagara. Following this, I turn to the methodology of the study, including details of the interviews and local press reporting that provide the basis for the analysis of borderline Canadianness.

Historical approaches to the study of borders serve to denaturalize extant state borders by demonstrating how borders have variously appeared and disappeared, as well as been contested and reworked through time. North American borders are largely the product of the bordered nation-state model that emerged in sixteenth- and seventeenth-century Europe and was then exported globally over the eighteenth and nineteenth centuries through an interconnected imperial world system of boundary making. As products of colonialism, North American borders are imposed upon coexisting national, subnational, and transnational polities. The violence of settler colonial bordering, in particular, continues to eliminate, displace, and contain Indigenous peoples, whose ongoing resistance in turn reveals alternative definitions and practices of territory, citizenship, and trade that challenge this violence (Borrows 1997; Grinde 2002; Simpson 2008, 2014).[1]

The Canadian government's official ceremonial motto refers to a nation that runs "from sea to sea" (often paraphrased as "from coast to coast"), highlighting the Atlantic and Pacific coastal boundaries.

Similar to the "Atlantic-centered migration narrative" described for the U.S. (Faires 2013, 39), however, it is more commonly the eastern coast of Canada that is featured as the most significant historical entry point for nation-building settlers from Europe (despite other eastward, northward, and southward movements), a narrative that is materialized in the Canadian Museum of Immigration at Pier 21, located in the eastern seaport of Halifax, Nova Scotia. The western coastal boundary, by contrast, has been more ambivalently incorporated into dominant nation-building narratives, being often identified as problematically permeable to migrants from Asia.[2]

Since the 2000s, political rhetoric in Canada has unofficially expanded to include the third coastal boundary of the Arctic Ocean, as is evidenced in the increasingly common phrase from "sea to sea to sea" (or "coast to coast to coast"). The Arctic boundary, already militarized during the Cold War, is receiving increased attention from the Canadian government as well as other Arctic states, as climate change increases the possibility of year-round ice-free waters and therefore the potential of greater commercial shipping and resource exploitation. The outcome is that this boundary is becoming more salient in the everyday nationalism of southern Canadians.[3]

Canada/U.S. Border Making

Although elided by Canada's "sea to sea to sea" rhetoric, Canadian land borders with the U.S. have also been central to Canadian nation-building and nationalist ideologies.

The Canada/U.S. land border, excluding the portion that marks the boundary with Alaska, is commonly referred to as the "southern" border in Canada and the "northern" border in the U.S., despite the fact that significant portions, including Niagara, are aligned east-west.

The history of this borderline reveals its colonial imposition on existing Indigenous polities through the 1713 Treaty of Utrecht between Britain and France, the 1783 Treaty of Paris between the British and the U.S., and subsequent agreements between Britain and the U.S. – the 1818 Convention of Commerce and the 1846 Treaty of Oregon (Sadowski-Smith 2014, 23; see also Carroll 2001). While the boundary that emerged over the eighteenth and early nineteenth centuries saw limited regulation, the U.S. and Britain did establish a small number of ports of entry for customs purposes (including at Niagara in 1801). There were also attempts on the part of both the U.S. and Canadian authorities to impose restrictions on ongoing Indigenous cross-border mobilities from the 1860s onward (Sadowski-Smith 2014, 23–4).

The first Canadian Immigration Act of 1869 did not apply to the U.S. border (Anderson 2013, 38–41), but the anti-Asian U.S. 1882 Chinese Exclusion Act and Canadian 1885 Chinese Immigration Act marked an expansion of Canada/U.S. land border control to include immigration enforcement that was followed by further measures aimed at restricting Chinese as well as Japanese and South Asian migration to and through the U.S. and Canada.[4] During the 1880s and 1890s, increased U.S. concern about unregulated European entries to the U.S. through the Canada/U.S. land border also produced the so-called Canadian Agreement of 1894 that required steamship and rail companies transporting migrants inland from Canadian seaports to allow U.S. prescreening of those in transit to the U.S. The perceived ineffectiveness of the latter, however, returned enforcement efforts to the land border. By 1907, forty-three U.S. ports of entry were involved in inspecting non-nationals, and in 1910, when a new Canadian Immigration Act increased inspection of an expanded category of "undesirables" arriving from the U.S., 175 immigration officials were working at 107 border stations on the Canadian side (Ngai 2004, 64; Klug 2010, 397; Anderson 2013, 72). According to Sadowski-Smith (2014, 25), by 1920 a newly "firm" Canada/U.S. border was rerouting U.S.-bound unauthorized migration to the then less regulated Mexico/U.S. border.

While greater regulation reduced what had been relatively "free circulation" of Canadian and U.S. citizens across the shared border, the primary targets of the U.S. Border Patrol that began policing between Canada/U.S. ports of entry in 1924 remained third-party nationals (Ramirez 2001, 45–7; Klug 2010, 397–8). Canadians (and Mexicans) were exempt from visa or passport requirements and were not included in U.S. quota laws that controlled entries based on national origins. In turn, U.S. nationals were not included in Canadian border entry requirements for visas and passports (Anderson 2013, 101–2).

Sadowski-Smith (2014) notes that the increased U.S. border enforcement at the northern border was the model for a subsequent hardening of the southern border – what she terms a "Canadianization" of the Mexico/U.S. border – well before what Andreas termed the "Mexicanization" of the Canadian/U.S. border in the post-9/11 period (Sadowski-Smith 2014, 30). Interconnections between Canada/U.S. and Mexico/U.S. bordering then have a long history that preceded the trilateral North American Free Trade Agreement of 1994 (and the subsequent, interrelated bilateral Mexico/U.S. and Canada/U.S. Beyond the Border initiatives, which occurred after my research period).

While it is important to recognize that interconnected North American bordering is disproportionately shaped by U.S. power, there are also important differences, given the contrasting economic and geopolitical positionings of Canada and Mexico vis-à-vis the U.S. When the northern border is targeted for greater securitization by the U.S., moreover, this has more often been framed as necessary for protection against the possibility of non-Canadian terrorists using Canada as a transit country rather than for controlling Canadian migration. In contrast to the U.S. constructions of the "southern" Mexico/U.S. border as a "cultural and racial boundary … [and] creator of illegal immigration" (Ngai 2004, 66–7), Canadian migrants in the U.S. are rarely collectively categorized as "illegals," or even "immigrants" (Ramirez 2001, 183; Bukowczyk et al. 2005). On the Canadian side, there was legislative exclusion of Asian-Americans and bureaucratic exclusions of African-American migrants in the early twentieth century (Shepard 1991), but white American immigration through the "southern" border has also remained largely unproblematized.

De-bordering/Re-bordering: Free Trade and Post-9/11 Securitization

The more recent history of Canada/U.S. bordering has been shaped by the bilateral and then trilateral Free Trade Agreements of 1989 and 1994, followed by post-9/11 securitization. The free trade agreements aimed to provide greater permeability for a broader array of continentalized flows and were considered by many to be part of a wider project of North Americanization premised on the concept of "North America as a meaningful political entity, economic arrangement, and cultural idea" (Ayres and MacDonald 2012, 22).

Much of the scholarly work on North American continentalism as a political project, however, highlights instead the tensions embedded in the unequal dynamics of the "asymmetric interdependence" (Andreas 2005) of Canada/U.S. bilateralism and/or "constrained hegemonification" (Clarkson 2006, 604) of North American trilateralism, wherein Canada and Mexico, dependent on trade with the U.S., consistently acquiesce to U.S.-led border and other priorities

Despite these tensions, continentalist visions remain part of elite border-related discourses in North America, and often shape contemporary government-funded and/or commercially commissioned border-related research. An example can be seen in the case of the Borderlands Project, based at the Canadian-American Center at the University of Maine and launched in the aftermath of negotiations over the 1989 Canada-U.S. Free Trade Agreement. Funded by the Canadian federal government, the provincial governments of Quebec and Saskatchewan, and the U.S.-based William H. Donner Foundation, it produced considerable interdisciplinary border-related research, including work on the Canadian and U.S. meanings of Niagara (McGreevy 1991). The Borderlands Project was described by its framers as "neither nationalist nor continentalist" but was explicitly based on the premise that "North America runs more naturally north and south than east and west as specified by national boundaries, and that modern communication and efficient transportation help to blur the distinctions between regional neighbours" (McGreevy 1991, iii). This premise was congruent with, if not determined by, the ascendant continentalist perspective on the Canada/U.S. border that accompanied the Free Trade agreement.[5]

The Free Trade context of continentalism was, however, altered by post-9/11 U.S.-led North American re-bordering, as both the Canadian and Mexican borders were identified as "weak links" in U.S. domestic security. The 2002 creation of the U.S. Department of Homeland Security signalled the heightened priority placed on securitizing U.S. territorial borders (Nicol 2011, 264; Salter and Piche 2011).

As border securitization became more significant to U.S. discourse and policy, Canada – despite its initially greater preoccupation with facilitating cross-border trade while maintaining the territorial border as "a line of defence" against U.S. economic, political, and cultural domination (Nicol 2005, 778) – quickly joined the new project of "securing flows" through the bilateral Canada/U.S. Smart Border Agreement of late 2001. This agreement increased bilateral cooperation and investment in border surveillance and technologies, with the result that Canadian border officials "now carry guns, deploy multiple means of surveillance, have access to, and share, U.S. databases, and work in integrated border enforcement teams (IBETS) often at joint border posts" (Nicol 2011, 273). In late 2003, the creation of the Canadian Border Services Agency reflected the new prioritizing of border securitization by the Canadian government. While some scholars described the new measures as "conspicuous consumption of policing" for a U.S. audience rather than effective security (Salter 2007, 312), the measures nonetheless had direct effects, including increased ethnoracialized and other forms of filtering at Canadian land and airport borders. Those seeking entry (and re-entry) to Canada at land ports of entry, for example, have been shown to be subjected to informal "discretion" on the part of front-line Canadian border officers, who differentiate border crossers on the basis of perceived attributes of race, nationality, ethnicity, culture, and religion while invoking national security as a rationale for such practices (Pratt and Thompson 2008). Work on extraterritorial post-9/11 Canadian bordering through varied forms of "remote control" also points to filtered im/mobilities and the significance of these for the re/production of im/mobilized inequality (see Mountz 2010; Razack 2010).

If border-related scholarship of the free trade era reflected powerful interests in North American continentalism, the post-9/11 period has seen applied border-related research focused on how best to combine cross-border commercial flows with securitization. From this perspective, Canada/U.S. border regions are framed as "corridors" or "gateways" for both lucrative (trade) but also dangerous (terrorist) cross-border mobilities. Some smaller studies have also more specifically assessed border resident reception of border securitization projects for the purpose of improving their efficacy (e.g., Ziolkowski 2006; Paulus and Asgary 2010).[6]

Canadian Niagara

Niagara is a significant site in the early history of First Nation–Crown relations as it was the site of a gathering between over twenty-four Indigenous Nations and British authorities that resulted in the 1764 Treaty of Niagara (Borrows 1997). The subsequent 1783 Treaty of Paris that had ended the war between Britain and the American colonies also established the Niagara River as the boundary between Upper Canada and the United States, thereby imposing a contested colonial borderscape on First Nations and others living on both sides of the river. Free movement and exemption from duty at the boundary line for "Indians" was established under the 1794 Jay Treaty, but uneven fulfilment of these and other treaty rights has been highlighted by long-standing Indigenous activism in the region. A 1927 meeting of Haudenosaunee leaders at the Peace Bridge was part of broader protest against colonial bordering, and annual border crossing parades since that time have continued to assert Jay Treaty rights (Grinde 2002, 177–8; Dirmeitis

2012). In 1993, the Seneca Nation unsuccessfully launched proceedings in U.S. courts to reclaim Grand Island and several smaller islands in the Niagara River (the case reached the U.S. Supreme Court in 2006 but was not heard).[7]

Imposition of British border regulation at Niagara began in 1801 when Fort Erie, Niagara Falls, and Queenston became three of eleven newly established ports of entry at which customs were collected.[8] The Niagara boundary, however, would be challenged by border battles during the War of 1812, which devastated the river communities on both sides. Later cross-border attacks led by William Lyon Mackenzie during the 1837 Rebellion and then Fenian incursions in 1866 resulted in ongoing border militarization. British customs personnel who were involved in anti-Fenian surveillance in Niagara and other border areas were part of an "embryonic secret service" (Whitaker, Kealey, and Parnaby 2012, 22–33), and the remnants of forts, monuments, and battle sites that line the Niagara River are reminders of a contested border even as they are promoted as tourist sites.

While restricting immigration was not an initial focus of border control at Niagara, this would become more significant by the end of the nineteenth century, by which time there were large cross-border flows of both migrants and day trippers. In 1880, 12 per cent of the population of Buffalo was from Ontario, and Niagara was a major corridor for short-distance migration between Southern Ontario and New York State (Widdis 2010, 452, 488). By 1894, Canadian labour migration was significant enough for Buffalo labour leaders to blame day and seasonal workers from Canada "for depressed wages and poor working conditions in the city" (Siener 2008a, 57).

As central state regulation of cross-border flows expanded at the end of the nineteenth century, business elites on both side of the Niagara River expressed concern about the allegedly negative effects on the regional economy (Siener 2008a, 37). By this time, Niagara was established as global tourism destination, and tourists often crossed the local border in the course of their visit to the area (Dubinsky 1999; Siener 2008a). In addition to this, many Canadian and American border residents were accustomed to crossing "for business, work, or pleasure, or to spend time with relatives and friends" (Klug 2010, 396). In this context, Klug (2010) documents how the U.S. Immigration and Naturalization Service "faced potentially conflicting political demands for effective inspection of all aliens," on the one hand, and "a speedy and unobtrusive inspection process" on the other (399).

During the nineteenth century, an increasingly regulated Niagara border also produced new forms of illicit cross-border migration that included the Underground Railway, created to smuggle fugitive African-American slaves across the Niagara River from the U.S. to Canada. The sanctuary role played by Canada is celebrated in dominant Canadian national narratives, but the history of Canadian Niagara also includes slave holding, the selling of slaves across the Niagara River, as well as attempted deportations of African-Americans seeking freedom (Murray 2002; Cooper 2009; Wigmore 2011).[9]

By the later nineteenth century, the new anti-Asian migration controls mentioned earlier also led to the smuggling of legally excluded Chinese (some of whom were prior U.S. residents attempting to re-enter) across the Niagara River. The outbreak of the First World War led to an expansion of those smuggled across to include " inadmissible" Europeans seeking entry to the U.S.[10] Prominent and more ordinary border residents on both sides of the Niagara River were involved in the organized transportation of unauthorized migrants across the river (Siener 2008a, 47–56). Locals involved in this activity often had past or ongoing employment in government, law enforcement, or railway transport and benefited from the reluctance of some local judges to impose severe sentences on those who were caught (Siener 2008a, 58). As Murray (2002) has noted, many Canadian border residents of this period "refused to equate smuggling with crime" (193–4).

In the 1920s and 1930s, U.S. Prohibition redirected smugglers in Niagara to the more lucrative transport of alcohol across the river. This activity, however, met with intensified enforcement efforts as the new U.S. Border Patrol, established in 1924, policed the river in speedboats equipped with machine guns (Siener 2008a, 59). American pursuit of suspected smugglers across the river led to fatalities on both sides, and incursions into Canadian territorial waters and soil severely strained local and binational diplomatic relations (Siener 2008b, 433–5).[11]

Bilateral and local cross-border tensions over border enforcement contrasted with official speeches at the opening of the Peace Bridge in 1927, which emphasized harmonious cross-border relations and encouraged local urban planners and Chambers of Commerce on both sides of the river to dream of a single "international city" of Niagara (Siener 2008b, 430). Such visions were challenged by expanded U.S. border inspection requirements that were slowing cross-border flows, as "submitting all incoming travelers to even a quick primary inspection proved challenging" (Klug 2010, 400). Klug's (2010, 408) analysis of letters of complaint to the U.S. Immigration and Naturalization Service from Detroit and Buffalo between 1933 and 1941 reveals ongoing tension between border residents, tourists, and U.S. border enforcement. Her work suggests that U.S. inspectors responded to contradictory political pressures to both increase regulation and facilitate cross-border mobilities by using their discretion to facilitate the crossings of those most likely to complain about alleged mistreatment at the border – notably the wealthier and better connected (Klug 2010, 412).[12] As this suggests, border crossings were officially and unofficially differentiated through filtered bordering within the larger context of increasing regulation.

Articulation of an "international city" concept at the time of the Peace Bridge opening, and complaints from regional elites about new forms of border enforcement affecting tourist and local cross-border mobilities, can be understood as early expressions of what I call official cross-border regionalism characterized by efforts on the part of local elites on both sides of the river to develop and promote shared cross-border visions for regional development. Another early example of this kind of initiative was the unsuccessful pitch, in the mid-1940s, of political and business leaders from both sides of the boundary, proposing various islands in the Niagara River (first Goat, then Navy, then Grand Island), as sites for the new headquarters of the United Nations (Siener 2013).

The promotion of cross-border regionalism by some local Niagara elites has continued into the present and is often proffered as a solution to regional economic challenges. American Niagara has experienced economic decline, while on the Canadian side – despite its early political importance as the site of the first capital of Upper Canada (1792–7) and a longer period of vibrant industrialization in the nineteenth century and in the first part of the twentieth century – the region is politically and economically peripheralized. The late twentieth century saw significant deindustrialization, most apparent in lay-offs and closures in the manufacturing sector. Although the tourist industry has expanded, this economic sector is linked to lower-paid and often seasonal service-sector jobs and more precarious employment, which, combined with the retrenchment of the Canadian welfare state under neoliberal federal and provincial governance, has produced weak household incomes and youth outmigration (Binational Economic and Tourism Alliance 2011).

In response to these conditions, some regional Canadian-side business and political leaders then joined with their U.S.-side counterparts to argue that the pathway to greater prosperity required a more integrated Canada/U.S. Niagara borderland that would attract expanded trade and tourist flows (Meyers and Papademetriou 2001; Macpherson and McConnell 2007; Schneekloth and Shibley 2005). In this context, some decry the allegedly weak forms of binationality found in Niagara as compared to the Pacific Northwest Canada/U.S. borderland region referred to as Cascadia (Eagles 2010, 380; Brunet-Jailly 2007). Illustrative of such weakness, it is argued, are the recurrent bilateral tensions associated with changes in the Peace Bridge border infrastructure as well as the loss of binational financial support for a cross-border summer Friendship Festival in 2005 (Eagles 2010, 387–8).

Research informed by the vision of cross-border regionalism can be seen in a Binational Economic and Tourism Alliance–commissioned study that looked at the Niagara border ten years after 9/11. This report was primarily concerned with the issue of timely border crossings deemed essential to the region's ability to compete with other Canada/U.S. corridors for cross-border commercial flows. The report combined quantitative information on border crossings with qualitative research based on "stakeholder consultation" that involved thirty interviews with those described as "project partners, key identified businesses, government agencies, and industry associations" (Binational Economic and Tourism Alliance 2011, 11). As this methodology reveals, the regionally initiated project was driven by, and more interested in, top-down preoccupations with, and experiences of, a changing Niagara border and border region. This study, by contrast, takes a bottom-up approach to border life in Canadian Niagara.

Research in Canadian Niagara

As indicated in the introduction, this research project on border life began in the era of the de-bordering processes associated with North American free trade agreements, and I designed an interview schedule aimed at eliciting retrospective accounts of local border crossings and discussion of nationalized and other border experiences and identities within this context (see the appendix). I began in the spring of 2001 with two pilot interviews with older colleagues who had lived in Canadian Niagara since childhood. These initial experiences gave me confidence in the interview schedule, and I started to recruit younger adults by posting signs in community spaces and (much more successfully) making brief presentations to classes at the regional university. I had conducted four more interviews with younger adults before the events of 9/11. While I carried on with some already scheduled interviews in the weeks and months following 9/11, as interviewees offered their more immediate experiences of changes at the local border (something easily accommodated by the open-ended interview questions), I made a decision to adjust to the unanticipated research realities by incorporating post-9/11 developments into the research project. The result was that I was able to record Canadian border residents' responses and reflections during a dramatic period of border life in Niagara. By August 2004, I had completed fifty-one interviews.[13] In addition to the interviews, I also conducted a systematic survey of the local press for local border-related reporting and the result was a database of over 1,000 local press reports.[14]

Among the methodological and conceptual challenges of border studies is a tendency to frame research in a way that "sees like the state" (Wilson and Donnan 2012a, 21). The challenge is considerable, given that the meanings attributed to any particular border will tend to be "closely related to the ideological state apparatus, ideological practices such as nationalism … and the material basis of such practices, which manifests itself in territoriality" (Paasi 2011, 14). In the case of this study, state power permeated the material that was gathered. Many of the interviewees, for example, had direct personal, family, or friendship ties to those employed within the local border infrastructure (see Heyman 2012, 55, for more on the prevalence of central government employment in borderland regions). Local press reporting also disproportionately highlighted and uncritically reproduced the perspectives of U.S. and Canadian officials on border-related topics.

Beyond these issues, my own research design reflected the strong pull of a methodological nationalism that assumes that "the nation/state/society is the natural social and political form of the modern world" (Wimmer and Glick Schiller 2002, 301). My initial interest in the border setting, for example, had posited a priori that it would be illuminating to study the everyday experiences of those living at a territorial border (see Green 2012, 579). At the same time, however, I attempted to work against a methodological *bi*nationalism that would limit discussion to the Canada/U.S. relationship alone by using the preamble of the interview schedule to locate the experience of "growing up in a border region" within a wider context of globalization beyond the U.S. Concerned that "nationally framed questions typically elicit nationally framed answers" (Fox and Miller Idriss 2008, 556), I also decided to place direct questions about national identity near the end of the interview in an attempt to avoid framing the discussion of border life and identity in nationalized terms from the outset (see the appendix). Despite these efforts, a major challenge throughout my analysis remained the need to maintain a critical lens on the nationally framed discourses of much of the interview and local press material.

Assumptions about the importance of the border for distinctly nationalized experiences and perceptions were also evident in the many conversations that I had with others about my research design. Most of the interviewees, as well as others with whom I talked about my study, asked if I was going to complement my Canadian-side work with documentation and analysis of the experience and perceptions of residents on the U.S. side of the border. As my research goal was to explore borderline Canadianness, I considered a study of borderline Americannness to represent a different project, but because my study involved asking questions about national identity – including about America and Americans (see the appendix) – my Canadian respondents often expressed interest in a "two-sided" study that they imagined might illuminate how "they" (Americans) saw "us" (Canadians). Significantly, it was an interest in this possibility of a mirroring back of perceived Canadianness rather than Americanness in and of itself (which many Canadian border residents felt that they already understood) that appeared to prompt the repeated question about my methodology. An apparent desire to see themselves through an American-side lens was itself revelatory of lived Canadianness at the border.

While my research was not set up as a comparative two-sided study, there are some examples of this kind of border-related research work at Niagara. The previously mentioned Borderlands Project, for example, included McGreevy's work on how the Niagara border region has occupied different positionings within Canadian and American nationalisms. For Canadians, he argued, Niagara serves as a celebrated national "front entrance," whereas for the Americans it marks the limits of American expansionism and therefore, he suggested occupied a more stigmatized national "back alley" positioning (McGreevy 1988, 1991, 1994).

Another small two-sided research study of local press reporting by papers in Niagara Falls, Ontario, and Niagara Falls, New York, over a two-week period in March 2001 noted that the Canadian-side *Niagara Falls Review* devoted much more (11.87 per cent) of its reporting to the topic of the border and the U.S. more broadly than the U.S.-side *Niagara*

Gazette did to the border and Canada (1.85 per cent) (Stieve 2005, 10). Evidence of such asymmetrical bilateralism at the Niagara border supports Gibbins's (1989) broader observation that "Canadians and Americans do not share the international border in the same way" (2).

As indicated above, because my primary research interest was in Canadian border residents' experiences and perceptions of the border, my interlocutors preoccupations did not lead me to alter my approach and I therefore retained a "one-sided" border research methodology with its resulting strengths and limitations. As it turned out, the recruitment process did produce some interviewees who had spent significant amounts of time living in the U.S., and some of their more U.S.-based insights are included in the analysis, but this was not my focus. Having outlined key elements of the methodology of the study, I turn now to share more details about the interviews and the local press material collected for this project.

The Interviews

My reliance on interviews and local press reports as the primary source material for a study of Canadian border life invites some discussion of the limits of these kinds of sources. In the case of interviews, for example, it is important to note that while border-related talk can be revelatory of "what borders mean to people" (Newman 2006, 154), such "border stories," as Berdahl (1999) notes, are ultimately "expressions and interpretations of lived experience, not necessarily depictions of actual reality" (173). A similar reminder of the limitations of interviewing also comes from scholars of everyday nationalism (Fox and Miller-Idriss 2008). Interviewing then offers narratives rather than direct access to everyday border life and of course, there may be a disjuncture between what people are willing to discuss about their experiences and perceptions in a research setting and their actual behaviours or beliefs. Because what is shared is further shaped by interview dynamics, I offer additional details about the interviewing context below.

I was the primary recruiter and interviewer for this project (I conducted forty-seven of the fifty-one interviews), and the fact that recruitment and interviewing were largely conducted in academic spaces meant that my older-adult, white-settler, female Canadian professor identity was always salient for the interviewees, who were predominantly younger university students. It is likely that my identifiers shaped recruitment as well as interview exchanges and silences. Two graduate students also assisted with recruitment and interviewing for the project. One of these, a female black Ghanaian student, conducted three of the interviews (with Anne, Jason, and Debbie) and was present at a fourth that I conducted (with Joe). The second student, who was female and Aboriginal, did some interviewee recruitment and also conducted one interview (with Paula). Again, it is likely that perceived identities shaped researcher-interviewee dynamics in these cases.

In an effort to partially unsettle unequal interview dynamics, I usually prefaced interviews by disclosing that I had not grown up in Canadian Niagara or any other border region and, as a result, was keen to learn more about border life from local "experts." Sometimes I also mentioned my mother's experiences of growing up in the border town of Sarnia, Ontario. For the most part, interviewees appeared at ease. Many commented that they found the interview experience to be an interesting one that made them reflect more deeply on border life and identity. Several volunteered that they were intrigued with the terminology of "border kid" that I had used in the course of recruitment and in the questionnaire that was distributed to interviewees ahead of the interview itself (see the appendix).

In my analysis, I include many direct quotes from both interviews and the local press in order to covey the richness, nuances, and complexities of border talk in this region. As the interviewees were offered anonymity, pseudonyms rather than real names are used, but other details of gender, generational cohort, border community and classed and/or ethnoracialized positionings are often included when the interviewee is first quoted. Below, I offer more details about the composition of the interview pool.

Of the fifty-one interviews, forty-four were with the young adults (twenty males and twenty-four females) explicitly targeted in recruitment materials. This group ranged from nineteen to twenty-nine years of age when they joined the study.[15] The remaining seven were older respondents (one male and six females), including the two pilot interviewees and others who responded to recruitment efforts. These older interviewees ranged in age from their thirties to sixties at the time of the interview.

Forty of the fifty-one interviewees had spent most or all of their childhood and teen years in the riverside Canadian border communities specifically named in recruitment materials (Fort Erie, Chippawa, Niagara Falls, Niagara on the Lake, and Queenston). All but one of the remainder had grown up within a fifteen-minute drive of the border in Canadian Niagara communities such as Welland, St. Catharines, and Thorold. Four of the respondents had spent some time residing on the U.S. side of the border. Jill had spent six months living there as an adult, while Chuck had moved to the U.S. side in grade eight and then returned to Canada after high school. Dylan and June were both born on the U.S. side but later moved to Canada, Dylan arriving as a preschooler and June as a young adult (making her the only interviewee who did not grow up on the Canadian side).[16]

Because I was interested in exploring how Niagara border life might be shaped by classed and ethnoracialized positionings, respondents were invited to outline the educational, occupational, and cultural backgrounds of their households of origin at the beginning of the interviews (see the appendix).

A minority of interviewees used explicitly class-based terminology such as "low income," "working class," "blue collar," "low to middle class," and "upper-middle class" to identify their families. For those who did not offer such descriptors, I relied on what they said about their parents' education and/or occupations to broadly assess their class backgrounds. In the end, I calculated that approximately twenty-seven grew up in working-class families (several of these emphasized their status as first-generation university students), an additional fourteen were from lower-middle-class households, and the remaining ten came from upper-middle-class backgrounds.[17]

Responses to questions about cultural background elicited varied responses, including some information about migration histories. As mentioned, two of the respondents, Dylan and June, were born in the U.S. and later moved to Canada. In addition to these, two of the older interviewees, Judith and Jill, had come to Canada as young children from Europe, and Joe and Ameena had arrived as youngsters from the Middle East and South Asia, respectively. An additional seven respondents identified at least one immigrant parent – all from Europe. Some mentioned parents who had come to Canadian Niagara from other parts of Canada, but twenty-eight identified at least one parent (and many had both) who had also grown up in Canadian Niagara.

Along with identifications of parents as Six Nations (Haudenosaunee) or French-Canadian, other ethnonational combinations and/or degrees of Canadianness were also referenced. Michael, for example, described his mother as "full Italian" in contrast to his father whom he described as "a mix of Norwegian and Belgian, but … more of a Canadian person." More distant and diverse migrant and non-migrant origins were highlighted in Dave's description of how his grandmother was "all sorts" because she was "British Isles, Irish, Scottish, French-Canadian, some Native, and some American." The parents of his Canadian-born grandfather, he also offered, were from "the part of Finland that's now Russia." Peter shared that one of his great-grandmothers was "from over in Europe somewhere," while Melissa described her family as "very Canadian," having arrived "many generations ago ... from England and Ireland." Chuck also described his origins as "Canadian, though originally half English, half Scottish."

Some interviewees contrastingly downplayed forms of migrant or ethnonational identification. Ed, whose mother had come as a child from England, described his grandparents as "Western European," but added, "If anybody asks 'what I am,' [I answer], 'Just Canadian.' I don't have any ties at all." And Katherine described her family as "English, so we really don't have a cultural background." Tessa answered the question about cultural background by saying "Nothing," and Dan described his family as "Canadian. Not a recognized minority or anything." Sheila, who claimed Canadianness, added "If you go back further, there's probably other cultures, but I don't know."

Chuck self-racialized as Caucasian, and a few others identified themselves as white in other parts of the interview, but the majority did not racially classify themselves. I used my reading of appearance and other interviewee statements (e.g., stated contrasts between themselves and racialized Others) to categorize forty-eight of the interviewees as white. The three remaining were Joe and Ameena (who, as mentioned, had migrated from the Middle East and South Asia, respectively), and Erin (who referenced both Six Nations and European-descended relatives).[18] While Erin referenced the Six Nations status of some of her relatives, she did not discuss her own status in this regard. Five of the interviewees identified themselves as dual Canadian/U.S. citizens, and a sixth was working towards this status. I have assumed that the remainder, most of whom did not volunteer their citizenship, were mononational Canadians.[19]

The Local Press

As previously mentioned, in addition to conducting interviews, I gathered over 1,000 border-related articles published in the *Niagara Falls Review*, a local daily paper based in Niagara Falls, Ontario. My focus was on reports related to local border-related events (rather than national newswire stories), letters to the editor, and editorials. I systematically reviewed the period from February 1999 (the beginning of the paper's online availability at the time) to late 2004 and then continued to gather material (but in a less comprehensive way) beyond that date. It was not always easy to determine which articles might be of most importance for my study, so the list of topics collected was flexible, and sometimes I went back to the database to search again for reporting related to issues that had emerged in the interviews. Of course, like interviews, local press reporting as a source of insight into border life has its limitations. The selection of stories and viewpoints are shaped by the profit-oriented logics of the newspaper business as well as journalistic constraints and conventions rather than scholarly preoccupations. With these constraints in mind, I use the *Niagara Falls Review* border-related reports to contextualize the interviews within local developments, to analyse reporting on a particular phenomenon (such as unauthorized crossings), and to illuminate official and everyday border talk and activities. Along with using local news reports as a source of information about a changing Canadian Niagara, I also use methods of critical discourse analysis to reveal how local coverage portrays some key border issues.

Conclusion

This chapter has provided a context for this study of border life in Canadian Niagara by first locating the Canada/U.S. border within a deeper history and wider political economy of North American bordering. After discussing something of the history of Canadian bordering, and more particularly Canada/U.S. bordering, I offered some details about the history of the Niagara border and its relationship to a changing regional political economy. The review emphasized how the Niagara border has been constructed and contested through time and how increasingly regulated border crossings were marked by forms of exclusion as well as filtered bordering. Some early examples of cross-border regionalism were noted and these provide some context for my later discussion of both nationalism and binationalism in the border region. Details of the research design and implementation, interviewee demographics, and local press reporting were also reviewed.

Chapter Two

Growing Up at the Borderline Pre-9/11

Introduction

As mentioned in the introduction to this book, when I moved to Canadian Niagara from Toronto, the salience of border talk in my early encounters reminded me of my mother's stories of growing up in a border community, and I wondered what it would mean for my own children to grow up in the Canadian Niagara borderland. At the time that I formulated this project, I was teaching in a Department of Child and Youth Studies (I later moved to sociology), where I focused on the many ways in which children and youth are deeply impacted by, but also directly engaged in, wider political economies. As a result, I had both personal and professional reasons for approaching the study of Canadian Niagara border life from the vantage point of young people. In my early exploration of comparative border studies, I had searched for and found some limited but intriguing scholarship on children, youth and borders (a field that has now greatly expanded).[1] When I applied for funding support for the project, I described it as focused on the experiences of young people in the border region, suggesting that listening to younger Canadian border residents who had grown up in the apparently de-bordering era of free trade would be particularly revelatory of this new reality and its possibly differentiated impacts on residents.

While I had begun the project highlighting the experience of growing up with de-bordering under free trade, the young people whom I was targeting for recruitment would be transformed into a cohort deeply marked by the events of 9/11 and subsequent border securitization. When it came time for the analysis of my interview material, it became clear that the latter had so shaped interviewee responses that I separated my analysis of retrospective accounts of the pre-9/11 period from that of the post-9/11 period – a separation reflected in the chapter divisions of this book. While this division risks uncritically reproducing dominant and popular claims that 9/11 "changed everything," it reflects interviewees' own portrayals of border life, even as their accounts reveal both change and continuity across this chronological break.

In this chapter, my analysis of the retrospective accounts of growing up in Canadian Niagara is limited to the accounts of the pre-9/11 period that were offered in response to a series of interview questions related to "crossing the border" (see the appendix). I reserve the accounts of the post-9/11 period for chapter 3.

The questions asked in the "crossing the border" section of the interview schedule reflected my initial sense of the significance of border crossing to everyday life in Canadian Niagara. As previously indicated, upon my arrival in Niagara, new neighbours and friends talked to me about cross-border shopping and recreational activities, and I learned from my students about other patterns of cross-border education and employment. It appeared that gathering narratives about border crossings would be a good way to begin exploring young border residents' experiences of the border and border life.

Responses from the predominately white, working- and middle-class interviewees drew upon their own experiences of border crossings as well as those of others in their social networks. While most of the narratives shared in the interviews related to being a border crosser, some respondents also offered understandings drawn from the perspective of border workers. Mary, Erin, and Lisa for example, had themselves been employed within the Canadian border infrastructure, while others knew family and friends who had been. The close connection of many interviewees to border-based employment was striking, and this connection clearly shaped their own experiences and perceptions of the border, filtered bordering processes and, to some degree, Canadianness.

As demonstrated below, the narratives of border crossing in the pre-9/11 period begin to reveal the significance of both cohort and age in experiences of border life. For the oldest interviewee, recollections of childhood border crossings went back to the 1950s, and this offered some intriguing contrasts with the experiences of the younger adults who made up the bulk of study participants. The childhood and teenaged years of the latter collectively spanned the 1980–2000 period, and within this range some suggested there were finer cohort-based differences when it came to border crossing, due to fluctuating exchange rates. The accounts also highlight life-course distinctions as interviewees contrasted their childhood, youth, and young adult experiences. These initial findings begin to illuminate differentiated experiences of border life and bordering.

I organize my discussion of the interview material by first focusing on retrospective descriptions of the cross-border activities associated with childhood and then turn to recollections of childhood experiences with the border inspection. I follow this with a discussion of the narratives of older teens' cross-border activities and encounters with border inspection. The accounts of both childhood and youthful crossings reveal the local border to be an everyday but also exceptional space traversed primarily for shopping, recreation, and the maintenance of cross-border kinship and other ties. The accounts reveal how, in the course of crossings, children and teens engaged with top-down bordering in varied ways ranging from acquiescence to active navigation to transgression of state power and regulation. The accounts also begin to illuminate local awareness of forms of filtered bordering, a topic that emerges here but is returned to in greater detail in chapter 4. Accounts of childhood and youthful border crossings also reveal how nationalized identities were re/produced in the course of what might be might be seen as a countervailing everyday binationalism. The discussion of border childhood and youth provides a rich entrée into border life and bordering in pre-9/11 Canadian Niagara while also offering crucial contextualization for the post-9/11 period.

Going "Over the River": Border Childhood

Interviewees' vivid accounts of childhood participation in adult-initiated border crossings reveal an "everyday geopolitics" in which "the negotiation of state power and sovereignty" can be a "simultaneously … intimate and state-making experience" (Greenberg 2011, 90). For some border children, the borderline itself, for example, could be made into an exciting part of childhood play. Peter (from a working-class family in Fort Erie) described this about his family:

> We used to cross the border all the time to … do our grocery shopping. I thought it was fun. Just staying in the middle [of the bridge]. Like being in the middle of Canada and the U.S. … that was always cool … my sister

and I ... she'd ride in the back and I'd ride in the front. I'd always yell to her, "I'm in the U.S. and you're still in Canada" and stuff like that.

As Peter's narrative suggests, for many children border crossings were part of an "everyday geoeconomics" insofar as these were linked to cross-border shopping. Because the frequency of this activity was shaped by exchange rates that shifted over time, different cohorts of border children were marked by differing degrees of involvement in this activity. Among the small group of older interviewees, for example, Nancy and Mary recalled how growing up working class in Niagara Falls, Ontario, in the 1950s and 1960s involved extensive childhood cross-border shopping, but that this became less common through the 1970s, when the purchasing power of the Canadian dollar was much reduced. Cross-border shopping intensified again from the late 1970s through to the early 1990s as the value of the Canadian dollar increased.[2] When the Canadian currency again lost value through the 1990s, cross-border shopping declined, picking up again in the post-research period as the Canadian dollar rose in the later 2000s. At the time of writing, in 2015, a lower exchange is again reducing the volume of cross-border shopping by Canadians.

The impact of shifting exchange rates for different cohorts of border children was highlighted by Judith from Chippewa, who pointed out how the return of more favourable rates for Canadians in the 1980s meant that her own children experienced the same regular "hauling over" to the U.S. side for groceries, gas, and running shoes that she had as a working-class child two decades prior. Brenda, from a working-class family in Niagara Falls, Ontario, also noted how as a child of the 1970s, she was much less familiar with cross-border shopping than older relatives who had grown up in the two previous decades.

While consistently highlighted by interviewees, exchange rates were not the only factor shaping cross-border shopping. Nancy, for example, recalled how the more prosperous Niagara Falls, New York, of the 1950s offered both cheaper prices and a wider array of products and leisure activities. The latter point was also highlighted by Mary's description of an early childhood crossing in the 1950s to see the Disney movie *Bambi* before it became available on the Canadian side. The younger cohort of interviewees, whose childhood crossings from the late 1970s to the early 1990s were often weekly or biweekly (although some went several times a week, while others only went a few times a year), also emphasized the greater range of shopping and recreational options on the U.S. side.

For working-class children, as I discuss further below, the major experience was one of shopping for basics. Paula from Niagara Falls, Ontario for example stated, "We used to go over all the time. We'd go grocery shopping ... It was cheaper for us to shop over there ... We'd go over and get bread and milk and groceries." Debbie, who grew up in Thorold, recalled how her aunt might hear of a special sale across the border and would call her family to ask if "anyone wanted to go shopping," and Kevin, a dual Canadian/U.S. citizen from Fort Erie, described how his family members would "go and gas up the car ... get groceries, maybe stop and buy some shoes, whatever you needed." Ed, from a working-class family in St. Catharines, related how his father did most of their grocery shopping "in the States," and "everybody went over to buy gas." Several others also described adult relatives enjoying particularly attractive prices by going to a gas station on an "Indian" reservation further inland on the U.S. side.

While children's border crossings were embedded within adult-initiated and -directed cross-border shopping, within these parameters, interviewees described forms of distinctly childish consumption. Mary noted, for example, how the promise of a Mars chocolate bar, unavailable on the Canadian side in the 1950s, was an effective "bribe" to secure her cooperation during adult-directed excursions. Among those recalling the late 1970s to early 1990s, Dan, from what he described as an "upper-middle-class area" in Niagara Falls, Ontario, recalled, "When I was younger you couldn't get yoghurt-covered raisins here [in Canada], so that was the big thing [in the U.S.]," while Sheila, from a working-class family in Niagara Falls, described the "neat, different things there ... There was a big candy store at one of the malls ... just over the border ... The gum was different and [so was] the packaging on a KitKat [chocolate] bar." The thrill of differently packaged milk was mentioned by Miles, from a family of "blue-collar workers" in Niagara Falls, Ontario, who recounted how "the biggest thing ... was that ... they had milk in plastic jugs rather than bags ... [and] chocolate milk in the plastic jugs was the best thing in the world ... Canada didn't have that."

Dale, from working-class Niagara Falls, Ontario, looked forward to purchases of a "special kind of Captain Crunch [breakfast cereal] that had the crunch berries in it," while Jim, from a lower-middle-class background in Fort Erie, also loved the cereal aisle of a U.S.-side grocery store, where he could "see all the … cool stuff you see on [American] TV, that's not available at home."

Along with enjoying the appeal of different and differently presented products, many reminisced about the pleasure derived from being given freer rein to select their own "treats." The "really good candy aisle" that had "all bulk, self-serve candy" was fondly remembered by Amanda from Niagara Falls, Ontario, who also described how she "liked going to the dollar stores because … [a relative] would give [her] five dollars and [she'd] go to town on a bunch of little erasers and paper and stuff."

Many suggested that a combination of lower prices and wider selection took on more significance for them as they became older and more interested in purchasing particular brand-name shoes and clothes that offered status and prestige within their peer groups. Dan explained, "My mom would take me over and buy me a pair of Nike Air [shoes], and I'd come back and all the kids would be like 'Oh cool,' and then two years later they show up in Canada." Rachel, who came from a lower-middle-class family, commented, "I really liked the clothes that AIMS [U.S. clothing store chain] had … We came back with cool stuff," while working-class Kevin mentioned how across the border he could get "more of the popular clothes, more of the brand names" that were otherwise only available much farther inland in the malls of Toronto.

Some directly linked the savings generated by cross-border shopping for staples to an increase in more discretionary expenditures. Because the money was so "good" (i.e., the Canadian dollar was relatively high), Melissa, who spent much of her upper-middle-class childhood in Chippawa, reported that her family could spend the occasional weekend at a hotel in Buffalo and buy "treats" for her. Tessa, from a lower-middle-class background in Port Colborne, recalled how every year on her birthday her aunt would take her across the border to eat out and to shop at the Toys "R" Us store, where she would buy "whatever I wanted."

For some families, cross-border shopping was combined with leisure activities. Mary recalled how the only restaurants that she visited as a child in the 1950s were on the U.S. side and that, because in the U.S. "kids could go to bars with their parents," she often accompanied her parents on cross-border evening jaunts if they couldn't find a babysitter. Many mentioned the pleasures of shopping and eating out. For example, Jessie, from an upper-middle-class household in Niagara

Falls, Ontario, recalled, "We'd call it 'going over the river.' We'd go to TOPS for groceries because the Canadian dollar was much better then. I remember going clothes shopping because Canadian money was at par. We'd go out for dinner there ... [We'd say], 'We're going over the river! It's fun!'"

Lisa, from a working-class background in Fort Erie, joked that "a major reason why we crossed [was] for chicken wings" – a treat that made the primary task of grocery shopping more appealing. Working-class Dylan from Welland described weekly cross-border shopping trips that included a stop at a buffet-style eatery, while Melissa's wealthier family had a "routine" of going first to a gas station, then to the grocery store, and then to a restaurant. Emma, from an upper-middle-class background in St. Catharines, reported that during the 1980s, "dollar-wise it was smarter ... to go over and do shopping there, so ... the whole family would go Friday nights ... for dinner and to do a little shopping."

As these reports begin to suggest, class positionings shaped and were shaped by border crossings in several ways. Those from more economically vulnerable households emphasized the importance of purchasing the basics of gas, groceries, clothes, and shoes more cheaply across the border. Emma, despite her more privileged background, also mentioned that after her parent's divorce, new economic pressures made it particularly necessary for her mother to buy butter, bread, milk, and gas in the U.S. Working-class Barb from Niagara Falls, Ontario, also recalled:

> My parents were just sort of struggling to come up to the middle-class standard. They wanted to provide us with the material goods that all the other kids had but [that her parents] really couldn't afford ... [It] was very beneficial for them to go over [the border] and take us to outlet malls ... to buy us clothes and shoes. We did that quite often.

Young people, then, were conscious of the relationship between exchange rates, border crossing, and their own household economies.

While less economically secure families sought to maximize purchasing power for necessities, young people in households with greater resources could engage in more discretionary purchases of "cool" American clothes, shoes, and other products, as well as the pleasures of a wider array of U.S.-based leisure activities. The intensity as well as the styles of cross-border consumption, then, signalled class positions within the everyday lives of Canadian border children.

As mentioned, cross-border shopping could be combined with recreational activities such as dining out. Some interviewees highlighted cross-border childhood leisure activities that included birthday parties at a U.S.-side children's entertainment centre, family trips to nearby U.S. amusement parks, and crossings to look at the Canadian Christmas light display from the American side. Several interviewees described childhood trips to Buffalo to attend major league hockey, football and baseball games and/or being spectators at sports events at U.S. colleges. Some participated themselves in cross-border children's sports leagues or competitions (e.g., soccer, lacrosse, dance). Others recalled crossing the local border as part of longer family driving vacations to the United States or to access cheaper and/or more convenient flights from the Buffalo airport.[3]

Beyond shopping and these kinds of recreational pursuits, some interviewees also described crossing the border as part of maintaining cross-border families, friendship networks, and cultural communities. Many made frequent visits to U.S.-based relatives or close friends, and the major urban centre of Buffalo offered important cultural community and religious activities that were accessed regularly by some.

Elementary- and secondary-level schooling could also be pursued "over the river." Jason, from an upper-middle-class family in Niagara-on-the-Lake, crossed daily to attend a U.S.-side school, while Nancy's childhood friend came in the other direction for Canadian education. Cross-border adult commuters from Canada were described as placing children in U.S. day-care centres, and Hannah, from a working-class family in Niagara-on-the Lake, made daily crossings in the summer to provide childcare to relatives living on the U.S. side.

Of these varied cross-border activities, shopping and recreation were the most frequently mentioned and were the most vulnerable to shifting exchange rates, which, as I have mentioned, shaped crossings in different ways for different cohorts. By the early to mid-1990s, as Canadian money became "bad" or "dropped off," many interviewees recalled reduced or eliminated cross-border trips. Barb, whose parents were struggling to achieve a middle-class lifestyle, recalled how during this period there was an "ongoing debate" as her parents calculated whether it was still worthwhile to cross over for particular purchases. By the time she was in grade nine, her family had stopped cross-border shopping altogether and – probably not coincidentally – this corresponded with when she began to feel pressure to get her first job: "My mother made me get my own job so she was no longer concerned with, you know, buying me brand-name clothes all the time ... She more or less said, 'You have a job, you have an income; you want these expensive clothes, you go buy them.'"

Jessie described how her family initially continued to vacation in the U.S., but it wasn't as enjoyable: "Because you think 'Oh, it's going to cost this,' and we have to think of the money much more, whereas when ... we were little and the money was better, it wasn't mentioned as much because it was so good." By the time she was in grades seven and eight, Jessie added, her parents were saying, "The Canadian dollar is so bad over there, it's not worth it anymore. We'll just stay here." Zak, from a working-class family in Queenston, explained how "the dollar started to drop" by the time he was in high school, and it was only worth going across the border for gas when they could maximize the more limited savings by filling the dual tanks in his father's truck as well as a "jerrycan for the lawnmower." Again, these narratives point to a high degree of awareness of, and engagement in, the vicissitudes of a regional border economy described as being shaped by see-saw exchange rates.

Interviewees also noted how the overall decline in the numbers of Canadian shoppers to the U.S. was accompanied by an increase in the numbers of American visitors to the Canadian side and the parallel expansion of retail and leisure services as well as service-sector employment. Many commented on how domestic shopping options improved through the 1990s, as more American retail chains "crossed" the border to set up Canadian outlets.[4] As Rob, from a well-off background in Fort Erie, stated, "In the past four or five years [1997–2001] we haven't been doing much [cross-border] shopping at all, because it's much cheaper to buy it in Canada. And the shopping experience in Canada has gone up because there's the new retailers, Commisso's, Sears, Wal-Mart, in Canada now."

As a result of these changes, Canadian border residents could access previously exclusively U.S. chain stores without crossing the border. The mobility of American capital then served to paradoxically "re-nationalize" consumption for Canadian border residents, as cross-border shopping became a less frequent experience when exchange rates were unfavourable. By the early 2000s, Jill from Niagara Falls, who was in her late thirties at the time of the interview, reported that if she and her husband took their kids over the border it was no longer for the "shopping or the groceries or ... even the gas" but for the less frequent activities of visiting friends, attending a concert, or embarking on a

U.S. vacation. Ed also noted at the time of his spring 2001 interview that "nobody ever goes over there to buy groceries anymore," contrasting this to times when his family "would go over once a week grocery shopping and get gas and then if we could convince my father to go to the factory outlet mall ... grab some T-shirts or some runners."

Dave, from an upper-middle-class background in Niagara Falls, who indicated early in the interview that his border crossing had been reduced as the Canadian dollar lost value to the U.S. dollar through the 1990s, later corrected this claim, noting that he had forgotten ongoing visits to relatives on the other side of the river. This "forgetting" revealed how family-oriented border crossing (in contrast to shopping trips) was not impacted by exchange rates and perhaps not always even experienced as border crossing at all.

While the frequency of cross-border shopping was consistently portrayed as resulting from pragmatic household-based economic cost-benefit calculations in a shifting border economy, discussion of fluctuating exchange rates was sometimes tinged with more nationalized discourses. When rates became less favourable for Canadians engaged in cross-border shopping, for example, Dan speculated that the higher U.S. prices (in terms of Canadian dollar purchasing power) represented a deliberate U.S.-side attempt to exploit Canadian visitors because he thought that "[Americans] got really used to Canadians coming over and ... would jack up their prices because they knew we were coming over and so our money wasn't worth much, and so it [cross-border shopping] kind of tapered off."

Jessie also recalled how her father, faced with less attractive exchange rates, offered an ex post facto protectionist argument for his reduced cross-border shopping, claiming that the U.S. didn't "deserve our money anymore if ... it's [the exchange rate] going to be so bad ... We can keep it in our own country and make ourselves profitable ... spending our own dollar here."

While interviewees were clear on the ways in which cross-border shopping could assist with ensuring household social reproduction for working- and lower-middle-class households while expanding the consumption and leisure options for the economically secure, the broader classed politics of a border economy were rarely referenced in the interviews. In particular, the ways in which the interests of more vulnerable border residents seeking lower prices across the river diverged from those of Canadian-side businesses trying to retain Canadian customers remained largely unaddressed. One of the few allusions to this came from Dale, who recalled the schools of Niagara Falls, Ontario, having a public speaking contest and that one student's speech about the negative impact of cross-border shopping on Canadian border businesses was televised locally. Cross-border shopping may have prompted "some resentment," among local Canadian business people, Mary acknowledged, but this was outweighed by "the practicalities of your life," because as "working-class people ... pressed for money ... if you could go over and get your school shoes and your back-to-school outfit for half the price ... you had to do that to survive." Mary's comment points to the challenges of Canadian-side business efforts to convince Canadians to adopt a more protectionist ethos. Nationalized appeals by Canadian-side businesses to Canadian-side shoppers to stay "home," were in any case replaced by marketing efforts aimed at enticing Americans to cross when exchange rates supported reverse flows to Canada. While cross-border mobilities were portrayed as a pragmatic way to maximize the purchasing power of household incomes in the context of shifting exchange rates, these coexisted with more nationalized framings that both paralleled and stood in tension with official cross-border regionalism.

The retrospective narratives of childhood crossings analysed here emphasize the role of macroeconomic exchange rates in shaping adult-directed cross-border shopping and leisure that shifted over time in intensity and purpose. Children from both economically precarious and more privileged households learned about not only the relationship between border crossing and their household class positioning but also cross-border consumption in the context of a deindustrializing regional economy. Uniquely "childish" forms of cross-border consumption were also linked to status within peer cultures. Outside of shopping and recreation, other forms of everyday binationalism included the maintenance of cross-border familial, social, and cultural ties less directly impacted by exchange rates.

Childhood and Border Inspection

Having conveyed something of the role of border crossing in "doing" border childhood in the pre-9/11 period, I turn now to more specific accounts of childhood encounters with state power as made manifest in border inspections that scrutinized the identities and intentions of border crossers. While childhood encounters with border inspections were mediated through adults, the narratives are revelatory of children's experiences, understandings, and sometimes more direct engagements with state bordering. They also begin to illuminate something of how children became aware of filtered bordering and nationalized identity in this context.

While the oldest respondents nostalgically recalled easy childhood strolls across the Whirlpool and Rainbow Bridges, the younger cohort emphasized varied degrees of childhood anxiety associated with entry into the U.S. Some mentioned how this anxiety could be exacerbated by adult efforts at coaching in preparation for the border inspection. Katherine, from a family she described as "low income" from Fenwick, described being woken up just before the border and instructed, "Be quiet ... keep your hands on your lap." Others were warned not to "joke around" or to be a "smart-ass." Kerrie recounted, "I was pretty scared when I was a kid because I remember my dad would undo all of the windows if we were in the back, so they could see how many people were in the car. And I remember he taught us to take off our sunglasses ... It was ... scary when I was younger." Removing sunglasses was mentioned by many others, including Anna from working-class Queenston, who stated, "When I was little, I would be wearing sunglasses ... in the summer and she'd [the adult driver] go like 'Take them off, we're coming to the border, that makes people think you're suspicious.'"

While some experienced heightened anxiety as a result of such coaching, in a context of frequent border crossing, others linked familial "rules" of border crossing with a reduction of their stress level. Kevin commented, "We had rules ... as we would pull up; you roll down your windows. If you have a hat on, take your hat off. If you had sunglasses on, you take your sunglasses off ... That's the way it was; it wasn't a big deal. We knew that's what we had to do."

The cohort differences between older and younger interviewees' recollections of border inspections reflected changes in border procedures on both the U.S. and Canadian sides, including changes in how child crossers were engaged. It was suggested, for example, that over time border officials became more likely to question children directly, as part of anti-child-abduction efforts. The role of border officials in retrieving missing children was highlighted in the 2001 bicentennial celebration of Canadian customs enforcement in Niagara, when the minister of Canada's Customs and Revenue Agency used his visit to the region to mark National Missing Children's Day ("Minister Praises Work of Customs Officials," *Niagara Falls Review*, 26 May 2001, A3). Several of the younger interviewees suggested that they were aware of the possibility of being mistaken for a missing child and of the need to demonstrate to U.S. officials that they were "properly" located children, travelling in the presumed safety of their own families.

Sometimes these procedures brought children into direct interaction with state power, when border guards required them to independently affirm their familial relationship to their adult travel companions. Several recalled being asked directly, "Are these your parents?" Ashley, a dual Canadian/U.S. citizen from an upper-middle-class household in Niagara Falls, Ontario, recalled how one time, when she was slow to respond to such direct questioning because she was half asleep, her parents became uneasy because her delayed response "looked a little strange." Chuck, who spent most of his teen years living in the U.S., recalled Canadian officials questioning him directly "because they were afraid … of kidnapping or some sort of situation where I was not related to my father, who was driving," and Rachel remembered, "My dad told me what to say because … if I just went with my dad, I'd have to say, 'This is my dad.'"

These experiences made the processes of filtered bordering apparent to even young children. That official constructions of vulnerable and domesticated childhood and the resulting regulation of child crossers could be linked to slowed or impeded cross-border family mobilities was revealed in Katherine's description of how her family car was pulled over when she was younger because she resembled a missing child and her adult relatives could not produce a birth certificate to prove her identity. June, an older, dual Canadian/U.S. citizen who spent her childhood living in the U.S., recalled how while attempting to cross together as teens, she and her older brother were refused entry to Canada: "I didn't have my birth certificate and they … weren't too sure of my brother's relationship to me."

Through direct engagements with border officials, presumptively vulnerable children were paradoxically placed in an agentive role as direct interlocutors when they were asked to independently declare or affirm their familial and/or citizenship status. Despite understanding that border officials were concerned with ensuring that they were in the care of "appropriate" adult figures, some recalled being apprehensive about the possibility of this kind of unmediated exchange. Melissa shared her anxiety about being told by her parents to roll down the window to say she was Canadian, while working-class Yvonne from Niagara Falls, Ontario, described nervously rehearsing her declaration of citizenship at the border: "I was terrified of it, actually. I remember as we were proceeding up to it [the border inspection], I kept asking my parents, 'What citizenship am I?' … I didn't want to screw up … I was really scared of that. So I always had to say, 'Canadian, Canadian, Canadian' in my head over and over again."

The unusual elicitation of younger children's voices could also, however, be recalled in more humorous ways. Zak, for example, recalled that despite the efforts of his mother to answer on behalf of his younger sister when a border official insisted on directly asking her where she was born, she "incorrectly" replied, "A hospital."

While the requirement to perform properly at the inspection was experienced as a daunting responsibility by some, the potential of a direct exchange with a border official could unsettle or even invert domestic adult-child hierarchies in ways that could also be exciting. Susan, who grew up in a lower-middle-class household in Niagara Falls, Ontario, conveyed something of these dynamics in her description of a memorable family crossing when an American border official directly addressed her:

> We only walked over the border once … Maybe it was just a nice evening … I must have been about seven or eight maybe and … we had to go inside the little booth and there was a flag on the wall and there was a picture of Ronald Reagan … The border guy talked to my parents, just the same stuff, you know, "Where you going? What are you doing?"… He didn't pay any attention to us because we were really little kids … and I remember at one point he looks at me and he said "Do you know who this man is?" and I looked at the picture and I said, "That's the president of the United States, Ronald Reagan." All of a sudden, he just lit right up from being so stern and mean (because I knew they carried guns in the States), to being so nice and he was just smiling and he was like, "Oh, what a good girl," you know … That's the only time anything like that's ever happened and it was just weird … I remember thinking when I was little, I wonder what would have happened if I'd said "I don't know," because I knew they had guns and things. Keep me there or something? I don't know.

This interaction remained salient for Susan, linked in her mind to an unusual degree of agency in an exchange with a powerful male adult American who elicited and then praised her "unchildlike" (and perhaps particularly "ungirllike") knowledge of U.S. politics. Child agency in this unequally structured interaction was both fraught (because of the possibility – in her own mind, at least – that a misstep might impede the cross-border mobility of her entire family), and empowering (insofar as she felt that her response was linked to the facilitation of their cross-border passage).

The sense of engagement and even empowerment was also apparent in narratives that portrayed the possibility of a direct interaction with border officials as adding piquancy to the experience of border crossings. Dan, who described his father as saying, "Sit there and be quiet and don't say anything," nonetheless found the inspection "kind of an anxious time, but also fun ... kind of a novelty." Jessie also suggested that she and her sisters enjoyed talking to the border personnel because "you got to know the border guards or whatever, and they'd just be like, 'How's it going? Citizenship?' We liked saying 'Canadian!' We got asked a question, we'd feel important or special ... we'd get excited to say it." Likewise, Tom from a working-class household in Niagara Falls, Ontario, described how "you felt kind of special in your own little way because you were young and you felt 'Hey, this guy cares what I have to say,' so you felt kind of special, but your parents always made sure to let you know to always be serious, no joking around, be straightforward with these guys."

While this latter account reveals the limits of child agency and, as indicated above, inspection was fraught for many, most of these predominantly white working- and middle-class interviewees emphasized the overall ease of pre-9/11 border inspections. Rachel, for example, stated that in the pre-9/11 period "they were very lenient ... It was just kind of like, 'Where are you from? Where are you going? What are you doing?' It was very quick, you'd go through." While there might be an occasional "cranky" border official, said Anna, "it was [usually] ... just, 'Hi, how's it going? Good. Okay, go ahead.'" While officials asked her directly about her Canadian citizenship when she was a child, she thought that this was only because "they thought I was cute."

Border inspections as experienced by children – whether recalled as more or less onerous – were understood to involve filtering processes linked to official constructions of children as being "at risk" in border space. In addition, many interviewees made comments that suggested an awareness of how such constructions and the resulting filtered processes could be gendered. For example, Michelle, from an upper-middle-class background in Welland, attributed the hassle experienced during a crossing with her uncle and father to the fact that "it was two males and a younger girl in the back. I guess it just didn't look right." The U.S. border official, she reported, asked her directly about her age and citizenship before pulling the car over for further examination. References to gender (and, as we shall see in chapter 4, to ethnoracialized and other identities) further suggest an early awareness among border children of filtering processes that facilitated some crossers while constraining others.

In the context of filtered bordering, children were instructed on the importance of performing at the border inspection "based on expectations of the border guard's expectations" (Salter in Johnson et al. 2011, 66), notably by demonstrating acquiescence to border authority and communicating what Salter terms "safe citizen" status. The need to convey acquiescence was highlighted in particular for U.S. entries, which were recalled as more intimidating in ways that linked the dynamics of the border inspection to the wider asymmetries of Canada/U.S. bilateralism. Several interviewees mentioned, for example, that they had found armed American border personnel particularly disconcerting. Rob commented that "all the customs agents on the American side [had] big guns, which is a little bit intimidating if you're a little kid," while Debbie added that, as a child, she was "scared of the border officials because I always thought, you know, we are trying to get into America so they'd arrest us because we were Canadian."

In the case of re-entries to Canada, there were also reasons for the performance of safe citizenship in the context of anticipated searches for undeclared goods. While the topic of smuggling was clearly potentially sensitive, the issue was raised with some consistency in the interviews. Although no one mentioned direct connections to organized smuggling, several offered descriptions of childhood involvement in adult transgression of Canadian border regulation through smaller-scale importation of goods for household consumption without payment of duty or tax upon re-entry. The degree to which interviewees mentioned their childhood involvement in this activity suggested that it was to some degree socially and morally acceptable within the region. Nancy was the most direct in this regard when she suggested that this activity was not considered "criminal" by locals, because undeclared goods had, after all, been paid for rather than stolen. Similarly, in work among border residents at the Mexico-Guatemala border, Galemba found that smuggling was "socially legitimized" as the idea of "economic justice" (2012, 718) was "detached from state-centered dichotomies of legality/illegality" (2013, 275).

Indeed in their accounts of re-entry to Canada, interviewees indicated an early awareness of tension between the top-down attempt by the Canadian state, embodied in border officials tasked with regulating cross-border purchases, and the bottom-up efforts of their adult travel companions to avoid such state regulation. As a result of this tension while re-entries to Canada were in some ways less intimidating than entries to the U.S., they were nonetheless fraught with the ever-present possibility of returning vehicles being "torn apart" by Canadian customs officials searching for undeclared purchases, given that, as Nancy put it, "everyone" experienced searches "at some point."

Even children in families that complied fully with Canadian state regulation could be apprehensive of such searches. Mary, for example, explained that her family had an "inflexible rule" against smuggling for fear of jeopardizing her father's employment as a customs officer, and Jill emphasized that her family was very careful about U.S. purchases because they were "worried about getting stopped." Many others, however, recounted how their adult travelling companions ignored such "rules" and recalled being directly implicated as children in adult subterfuge, insofar as they remained silent and did not contest adult misrepresentations to border officials. Tom told of his mother going over for groceries and gas but then declaring only the groceries while explicitly telling him not to mention the gas. Rachel described how "a couple of times we put a few bags under the seats because we had spent a little too much … [and were told by adult relatives] 'Shhh! We do not have anything under the seats!'" Lisa from Fort Erie recalled her mother telling a U.S. border official that the family was going to a nearby resort, when they were in fact only crossing to shop at the duty-free store.

Some respondents emphasized their comfort with such deception. Paula stated, "We'd get running shoes … throw our old ones out [and then] smile coming across the border. It's like 'No, I'm, not wearing bright new white runners!' … It was pretty easy." Others however, found the need for collusion to be stress inducing. Ed, whose father would hide alcohol "under a wheel well, or … seat, somewhere that it wasn't going to roll around and make any noise," recalled always having "butterflies" in his stomach when re-entering Canada, because, as he said, "Every time you came over, you knew you'd exceeded the [duty-free purchase] limit." Brenda, who as a child of the 1970s, had less experience with cross-border shopping and, therefore, less experience with minor smuggling, described her fear that she and her relatives might "be caught and put in jail" when an older relative told one of the other children in the car to hide newly purchased shoes by keeping her feet under the car seat during a re-entry to Canada.

As this anecdote reveals, along with remaining silent or not contradicting adults, children could be more directly implicated in concealing undeclared goods. Emma reported, "My parents used to bring things back illegally ... [and] we were actually sitting on the cases of pop on the seats in the car and pretending that we weren't." Zak described his parents placing a case of beer or bottles of vodka into his hockey pants for the return to Canada after a tournament, while Dale described a more complex system of evasion, whereby his parents would give him money. "I would have a set of groceries for myself ... They would divide the expenses between people to avoid having to pay duty on things ... I knew we were beating the system ... I mean, they'd certainly never ask me to do that in the grocery store down the street." As Ed recalled, his father would

> go to the grocery store and ... lay out ... what he thought totalled around twenty bucks and he would pay for that, and then he would have a second lot of groceries he would pay for again, and that ... receipt just got torn up and thrown in the garbage, so when you pulled up to the border you could produce a legitimate receipt and say, "I bought twenty dollars' worth."

Smuggling by adult family members and other acquaintances then implicated children from a young age in these transgressions of Canadian state power and vested them with a degree of responsibility for the success of this activity. In some cases, respondents bemusedly described how, as younger children, they were not sufficiently savvy to their important role. Kerrie recalled:

> I remember my dad, when I was like four or five ... would buy horseradish over there [in the U.S.] ... We were crossing once, and the customs officer asked if we had something to declare and Dad said no. And I go, "But the horseradish you hid under your seat!" And my dad was like, "Oh, yeah." So that was that. The guy [customs official] just laughed at us.

As this story suggests, however, part of socialization into border culture was learning that Canadian border officials didn't necessarily pursue investigation even when subterfuge was revealed. While Canadians were supposed to pay duty on taxable items if they had been out of the country for less than twenty-four hours, interviewees knew from experience that this state regulation was enforced only intermittently at the local border where, as Susan claimed, "No one really cared [if] you had an extra twenty dollars' worth of groceries."

The latter point was particularly clear to children who knew the local border as a place of employment for family, friends, and friends of friends. Personal connections to the border infrastructure produced a sense of "backstage" familiarity with border inspection processes. Kevin, who had relatives working at Canadian Customs, for example, recalled, "Coming home sometimes, we would know what lane they'd be working in, and we'd go through that lane so we could stop and talk to them ... [It was a] familiar, friendly setting that was never a problem."

Thus, while Mary emphasized that her father's employment as a customs officer had precluded minor smuggling by the rest of her family, others described how their household members' personal connections to border workers sometimes facilitated familial re-entries with undeclared goods. Kerrie, who told the horseradish story cited above, for example, also said that her mother was "good friends with somebody who worked in customs, so he would come [cross-border] shopping with us, and when we came back, we really didn't declare much."

As border residents, many felt that they were more privy to "backstage" border operations than outsiders to the region. Consistent with this sense of partial insider positioning, those who had actually undergone searches that resulted in family members or friends having to pay duty on undeclared goods presented these experiences as both unusual and unreasonable. Richard from Niagara Falls, for instance, recalled how surprised he was one time when he was returning from a dinner on the U.S. side with his friend's family and Canadian customs officials "tore the car apart," confiscating cigarettes bought at the duty-free store and impounding the car.

Some of the interviewees who shared such incidents portrayed them as unnecessarily harsh and arbitrary applications of Canadian state power, while also highlighting how experienced local border crossers could still negotiate their way out of these situations. Anna recalled, for example, that when she was young her family was "pulled over" upon their return from a winter family dinner at a U.S. restaurant. After an hour of being searched, she said, "everyone had had enough," so her mother "got out and yelled at the officer to let us go back [to Canada] and it worked." There was, then, a degree of entitlement made evident in these accounts, linked to not only the idea that re-entry to one's home country should not be denied but also the expectation that local identity should be associated with particularly easeful returns.

In the course of frequent local border crossings, children experienced both U.S. and Canadian state power at the border through both adult-mediated and more direct encounters with border officials in the course of border inspection. Through such experiences, they learned the importance of performing a childish version of "safe citizenship," notably domesticated deference and passivity, as well as officially elicited agency. Canadian children involved in minor inbound smuggling were also co-opted into adult transgression of Canadian state power. While such involvement revealed to children a tension between central Canadian state power and the everyday border practices of family and friends, many also learned that when this power was embodied in known individuals, these could be understood as likely (but not guaranteed) to use their discretion to facilitate such activity. As Galemba (2013, 279) points out, insofar as smuggling can reduce "regional discontent" linked to economic pressures, weak enforcement may be important to state efforts to ensure the support of border citizenry – a point relevant to economically peripheralized Canadian Niagara. The retrospective accounts of childhood encounters with both U.S. and Canadian border inspection, then, revealed varied indirect and direct forms of engagement with top-down bordering, and begin to suggest how such encounters also led to early awareness of both filtered bordering and nationalized identities.

Border Crossing and Youth

While the accounts of childhood crossings involved recollecting a more distant past, stories about teenage border crossings were often relayed with greater immediacy and were indeed often the most animated part of the interviews, despite the fact that for many of the younger cohort, a weakening Canadian dollar reduced cross-border shopping during their teenage years.

Despite less favourable exchange rates that reduced cross-border shopping, crossings for this purpose still continued. Kerrie from Welland described how, even with less favourable exchange rates, it could still make financial sense to access bargains online and have products shipped for free to U.S. friends or relatives and then cross over to pick them up. Others commented that shopping at the mall right across the border also remained more accessible than at inland Canadian malls.

Some of the accounts point to class-based differences within the overall reduction in U.S.-bound cross-border shopping, as youth from better-off families could continue with this activity even when it was less economical. Anna, from a working-class family in Queenston, for example, shared the sense of disadvantage that she experienced when her wealthier peers would "still go over there [to the U.S. side] and [they'd be] like, 'Oh, I like the Old Navy over there so much better' … [but] I [couldn't] afford to be over there and to even pay for the money [exchange]." In a period of less economically beneficial exchange rates, U.S. purchases could be imbued with increased cachet. Joe from Fort Erie described how even when his family stopped cross-border shopping, he would still go to Buffalo for particular brands of clothes and shoes unavailable on the Canadian side "because [he] didn't want to have the same clothes that everyone else [had]." Such patterns then could exacerbate classed divisions among Canadian border youth.

In a study of rural American youth living at the Quebec-Vermont border, Dunkley (2004) describes how the U.S./Canada border crossing was a "gateway" to a Quebec bar scene for U.S. border teens. Similar to Dunkley's description, I found that a consistently mentioned youthful cross-border pursuit for Canadian Niagara interviewees was cross-border drinking. Changing state (provincial and state level) regulations shaped this activity. In the 1950s, when the legal drinking age was eighteen in New York State and twenty-one in Ontario, Nancy had her first alcoholic drink on the U.S. side. In the 1960s, Judith reported, "Everybody went over to the States on Friday and Saturday night," and according to Mary, speaking of the same time period, Canadian youth would frequent U.S. bars because "you couldn't drink in Canada … [where] the liquor laws were applied very seriously … You had to go to the U.S." On Friday nights, she reported, "we were like lemmings. We went to Niagara Falls, New York, to drink … Different schools went to different places, [and] different age groups had their different places," but "you had to contrive to get there one way or another."

While drinking ages on the two sides of the border shifted over time – in Ontario, it dropped to eighteen in 1971 and then increased to nineteen in 1978, while in New York it shifted upward to nineteen in 1982 and then to twenty-one in 1985 – Ed recalled how cross-border drinking remained a "thrilling" part of teen border life in the 1980s. While you needed to be older to drink in New York, he recalled, "all of us as teenagers could get … into the clubs" because they did not require photo identification for entry "only ... a birth certificate ... [which] you could borrow." Recalling the 1990s, Barb described how her peers "would hop in a car, drive over to the States ... to the bars ... Their [U.S.] drinking age is higher than ours, but apparently ... fake ID works a bit better over there."

Ed portrayed the "traffic jam" of teenagers returning to Canada in early morning hours as a "rowdy" scene: "Everybody would be honking their horns ... [and] these guys that were drunk ... [would be] out on the middle of the [border] bridge fighting ... [and because] the American cops don't come over [and] the Canadian cops don't go out there, it was just a free-for-all."

This portrayal of border youth engagement with the bridge as a liminal space unregulated by state power is a striking one, but Ed also recalled that youthful crossings were affected by forms of regulation, including a New York State law prohibiting those under eighteen from driving after dark. After media reports of a Canadian teen having to "push the car to the border," he recalled how Canadian parents became "terrified" about this possibility and, as a result, he and his sixteen-year-old friends had to walk rather than drive across. New York State loitering laws were also used against Canadian youth, he claimed, because after the bars closed, "if you didn't go to your car ... [but] stood around for five minutes ... some police officer came up to you ... and if you didn't leave, you went in the back of the paddy wagon."

As these narratives suggest, the experience of independent crossings with peers could be an exciting, if fraught, activity even outside of the activity of drinking. Richard recalled how he and his friends, aged seventeen, "just decided, 'Hey let's go ride across the border' ... [and] rode [their] bikes right across." Kevin, whose high school was close to the border, recounted how as newly licensed drivers, he and his peers would sometimes skip class: "[We'd] go over to Buffalo and have lunch ... maybe go catch a Bisons game and then come home in the afternoon ... Or we would ... gas up our cars ... go over there on a Thursday night and hang out with some [U.S.] friends." Peter also described participating in a "tradition" of older teens crossing the border to purchase burgers: "On the last day of school ...we'd always pack cars and ... race over there." These examples suggest that earlier childhood games based on the border bridge expanded into other forms of border "play" for youth in Canadian Niagara.

Other cross-border leisure activities for teens included watching or participating in organized sports events such as wrestling, lacrosse, cheerleading, and Nascar racing. Barb mentioned a church-based youth group renting a roller skating rink across the border, while other interviewees participated in weekly crossings with their Canadian high school ski clubs through the winter months. As those from wealthier families could more easily partake in such pursuits, involvement in these activities, along with shopping, marked and reinforced class status. Avery, from a working-class family in Niagara-on-the-Lake, described her sense of being left behind as her more economically privileged friends "were over there all the time … they'd get the ski pass to go … every weekend … just load up in a van and go over."

Dunkley's (2004) study emphasized how the ability to participate in Vermont-Quebec border crossing was unequally structured by age and gender; younger teens were less likely to have the requisite access to cars, and young women were often subject to a parental "culture of protection" that limited their participation. The structuring of differential access to cross-border pursuits, she noted, allowed older males to gain prestige from sharing "heroic narratives" of "drunken hilarity, new friendships and memorable road trips" (even while they experienced the sometimes fatal risks associated with drinking and driving) (573–4).

There were certainly parallels with Dunkley's (2004) examples in the stories told by Canadian Niagara male interviewees about their recreationally oriented, peer-organized crossings. There were also hints of gendered restrictions, as some female interviewees mentioned their parents being worried about their teenaged daughters crossing the river. Jessie described how her mother would be "anxious" that she and her female friends "would get hurt or someone would do something bad … so we didn't go over … Even now, she's like 'Uh, I don't like you going over there.'"

Barb, whose parents also restricted her ability to join in cross-border drinking, noted at the time of the spring 2004 interview that while some of her girlfriends were "thrilled" to resume cross-border shopping, she was still nervous about U.S.-side crime and felt that "a bunch of girls" could be vulnerable in "a different country." Susan also reported how her mother's insistence that she have health insurance coverage for any cross-border visit, limited her participation in peer culture because if her friends decided "on the spur of the moment to go over the river for something to eat, I always have to buy insurance and it's such a downer because everyone [else] just goes and I'm not allowed to." The significance of border crossings for youth, including those who were unable to participate, then remained high – though in ways that heightened classed and gendered inequalities during periods of less favourable exchange rates.

Dunkley (2004) described border crossings as being part of a "cultural rite of passage" to adulthood, and in Canadian Niagara there was a similar sense that such crossings, made independently of older adult supervision, marked a significant shift – in part because they required an autonomous navigation of the border inspection. For Jill, for example, a teenage re-entry to Canada was described as "nerve-wracking because I didn't have my parents … there for me," while Lisa recalled being "very nervous" during her first time crossing the border without family. For several, the experience of independent crossing also produced new understandings of the international boundary. Nancy, who crossed regularly in the late 1950s to visit her best friend in the U.S., recalled how surprised she was when a U.S. border official unexpectedly asked her for documentation. She described this as a "watershed" moment that made her realize that the border crossing was "not just a stroll across a bridge." Ed also recalled the shock of being hassled as a newly independent crosser:

> The first time I was walking across … they want to know your citizenship … then they want you to produce … something that proves you're a Canadian citizen, and I gave them my driver's licence and I get yelled at. They [the border official] said, "An Ontario driver's licence doesn't prove anything other than the fact that you're entitled to drive a car in the Province of Ontario. It doesn't prove citizenship." I had never thought of that before … I was just … sixteen … I didn't think of things that way. So then it was, "Well, who are you with? Why are you walking over here? Where are you going tonight? Why aren't you driving?"

Increased awareness of international boundary regulation was the result of youthful autonomy, but several suggested that it was also due to more onerous inspections for older teens travelling on their own or with peers. Hannah, for example, suggested that U.S. officials were "more sceptical of you when you come over. They ask you questions like 'Have you been arrested?' and stuff like that. I guess their assumption of teenagers [is that] … we're going to go over there and drink or party." She went on to describe an incident when she was pulled over and questioned by U.S. authorities for an hour, adding that this would never have happened if she had been with her mother.

Dale also described how both sides of the crossing gave youth a harder time. In the case of the U.S., he claimed, "If you were going over on a Thursday night ... [U.S. border officials] think you're going to be drinking, you're not going over there to fill up your car with gas ... They would pull us over and sometimes make us go into the customs office and ask ... to see ID." Upon re-entry to Canada, he added, they would be subject to officials applying the "third degree," sometimes using dogs to search for drugs. Kerrie added that when her group of friends would "cross for fun, late at night after the bar or something ... go over just to see what was going on," they would "always get pulled over [by Canadian officials]."

These accounts again revealed awareness of filtered bordering processes based on official constructions of youthful crossers as "risky" rather than the "at risk" crossers they had been as children. Despite this, most downplayed any difficulties and emphasized their skills of border navigation. Erin, who spent her working-class childhood in Fort Erie, stated, "When you are a teenager ... they do seem to ask a lot more questions," but claimed that while her friends would get nervous, she "would say, 'What's the big deal'? Just say 'We are going shopping!'" Peter also downplayed any hassle, commenting that border officials turned a blind eye to the "weekend ritual" of cross-border drinking by local youth: they "[could] tell where we were going and ... [that] we'd be gone for the night ... [and] after a few times they kind of knew us."

As in childhood, youthful crossings could be eased by personal connections to border workers – now sometimes peers. Lisa recounted that she had attended high school with some border guards and knew others from her own job at the bridge. She added that she became aware that her own eased crossings were unusual only when crossing with friends from outside the region, who were "shocked" at her starting a casual conversation with the border official – who was a high school acquaintance. She emphasized that a "loose" pre-9/11 border facilitated her peer group's regular convoy of cars heading to and from a U.S.-side burger outlet.

Interestingly, within this context more troubled border inspections could be described by interviewees as adding to the excitement of youthful border crossing. Tom for example, reported that because he had often wondered what happened during a secondary examination, when he was subjected to a full search of his car at the U.S. border while crossing for a concert in Buffalo, he merely thought, "Hey, this is pretty cool!" In a less dramatic example, Dylan's account of how his "huge obsession" with U.S.-side Krispy Kreme donuts took him across the border included a description of the fun involved in performing with self-conscious youthful frivolity for border officials: "We'd ... cross the border just for Krispy Kreme donuts and when he'd [the border official] say like 'What's your reason for crossing?' we're like, 'Donuts' ... The man just kind of looked at us like, 'Don't you have anything better to [do?]'"

Mary also described how more serious difficulties with entering the U.S. could be turned into a game by Canadian border youth. In the 1960s, she and her friends would try to deceive U.S. border officials – whom they perceived as trying to thwart Canadian youth crossings to U.S. bars – by pretending to be tourists crossing to look at the Falls from the U.S. side. If this subterfuge failed and they were still turned away by border officials, she explained, they would then drive down to another bridge to see if they "could sneak over that one ... sort of 'play bridge.'" Yet another way in which border inspection could be reworked into youthful antics was revealed by Peter's description of how he would tell Canadian border officials upon re-entry to be sure to check the car behind him for undeclared items – thereby ensuring that his friends in that car were given a "hard time."

As Peter's story suggests, youthful re-entries could also sometimes involve forms of minor smuggling that provided an added element of frisson to encounters with Canadian border officials. Ed, for example, recalled bringing in undeclared goods, wondering, "Are we going to get away with it again? ... [Maybe] this is the night we're going down," and, after getting through without incident, feeling "elation": "... like you just committed the biggest crime in the world ... You're running from the law, [and] you just got away with ... sixty dollars in clothes ... At that age, it was a big deal." Providing more details about how to avoid more thorough searches, he recalled one time when he was "pulled in" to secondary examination near Christmas with "a car load of stuff." Knowing that he was going to have to pay duty, his strategy was to "hop out before that person [the border official] got to [his] car":

> I'd flash them the piece of paper and say "I have to pay duty" and then throw a receipt down on top of it so I would look like I was willingly going through with it ... meanwhile you still have a couple hundred dollars of undeclared stuff in your car. They never even looked at because you just acted like a nice person ... you didn't want them to catch you with that

> other stuff because you'd probably have to pay fifty dollars in duty which as a teenager you didn't really have.

In this account, Ed emphasized his skills of border navigation, adding that he had learned not to "declare everything" and to head for the line where Canadian border workers appeared not to care and "were just zipping people through." Even if you were "hauled in," he noted, the penalties were pretty minor, because "if you had a tank of gas, twenty dollars in groceries, maybe a T-shirt," you would only have to pay duty and maybe receive a "hard look" from border authorities. Lisa also discovered from her many crossings that even if she was wearing several layers of newly bought U.S. clothes, this received no attention from Canadian border officials: "It got to the point where I would not wear my [undeclared newly purchased] clothes over. I would just throw them in the bags in the back." According to her, Canadian customs would ask, "What, are you bringing over?" and she'd reply, "Oh, nothing, like twenty dollars' worth of stuff. I went to my [U.S.-based] cousin's house," and the officials would respond, "A-okay."

Avery also talked about successful subversion of state border regulation when she told of a friend-of-a-friend who

> had a secret zipper thing in his shoe where he could hide marijuana. And he's like, "It's the coolest shoe, you've got to get it if you're going to be doing this" ... My one friend told me ... how they told him just to pull over to the side and ... said, "Open your trunk," and there was nothing in there. It was all in his shoe! So he didn't really get caught.

While none acknowledged involvement in larger-scale smuggling, Ed made a passing reference to an acquaintance whose unsuccessful attempt to return to Canada with a "large quantity of cigarettes" resulted in his car being impounded. An expectation of eased crossings meant that experiences of border hassle upon re-entry to Canada were often portrayed as anomalous. Jill, for example, recalled being "very upset" in her early twenties when a Canadian border guard "did a complete thorough check of our car. He even wanted to look at the bottoms of our boots, check out the shoes that we were wearing." At the time, she said, she was outraged and thought to herself, "This is supposed to be my country, and you're hassling me like this?"

It is hard to know how open or circumspect interviewees were being, but the frequency with which youthful involvement in minor smuggling was mentioned was striking. Only Ashley noticeably censored herself by truncating a story that she had started to tell about a crossing, saying, "I have to be careful what I say here." Joe was also unusual in sharing some moral qualms, noting that while he used to be angry if he was pulled over and had to "pay American tax and Canadian tax," he had begun to feel that he "should not do that kind of thing" as it was not "very truthful." Such comments contrasted with the more common celebration of becoming savvy crossers able to successfully evade Canadian state regulation at the border.

Conclusion

In this chapter, border crossing narratives from the interview material are used to illuminate how young border residents experienced and imagined the local border in the pre-9/11 period. For most of the interviewees, this was a period when the de-bordering logics of continental free trade were in ascendancy, but the accounts of childhood and youthful border crossings analysed here put greater emphasis on the significance of fluctuating exchange rates in determining the intensity of everyday binationalism for different generational cohorts. Different cohort experiences were described by interviewees as further stratified by class positionings associated with household vulnerability or security in the regional cross-border economy. More precariously positioned working-class and lower-middle-class households were portrayed as taking advantage of favourable exchange rates to purchase basics more cheaply, while those from more economically privileged households were presented as able to enjoy disproportionately greater access to a wider range of discretionary shopping and leisure options on the U.S. side, sometimes even when exchange rates offered less of an incentive for border crossing.

The accounts highlight how cohort and class were articulated according to age and life course, as interviewees distinguished between their childhood and teenage cross-border mobilities and encounters with state border inspection. Learning how to handle a variety of possible encounters with both U.S. and Canadian border officials was part of growing up at the border, and young people's engagements with state power in the course of border inspections were described as producing both anxiety and excitement, as young people learned to acquiesce to, navigate, as well as transgress such power in both childish and youthful performances of "safe citizenship." The interviews suggest that young people learned about generationed and classed, as well as gendered and other forms of filtered bordering (discussed in greater depth in chapter 4). The interview narratives also begin to illuminate how young people's nationalized identities were made salient and thereby re/produced though border crossings. By documenting and analysing young people's experiences and perceptions of border crossings in the pre-9/11 era, this chapter provides important context for accounts of post-9/11 developments.

Chapter Three

Experiencing 9/11 and Post-9/11 Securitization at the Borderline

Introduction

Like many other Canadians, I keenly recall the day of 9/11, during which I joined a group of faculty, staff, and students gathered in hushed silence around a hastily set up television at the university. I remember a palpable sense of fear and vulnerability fuelled by rumours that there were many more planes in the air and that some might be targeting nearby Canadian sites. It did not take long before it became clear that 9/11 and the subsequent U.S.-led border securitization that followed would lead my research project in unanticipated directions.

The six interviews that I conducted prior to 9/11 had gone well, but I embarked on the first few post-9/11 interviews with some apprehension, unsure of how interviewees might have been affected by the events that had shaken so many. The interview schedule, designed to focus on retrospective accounts of border life, had succeeded in eliciting the rich material discussed in the previous chapter, but questions about past experiences of border crossing and identity had new meaning in the post-9/11 context. I wondered whether I would have to rework my interview schedule in response to the new research realities, but as the interviews proceeded, I realized that my open-ended, semi-structured interview method was flexible enough to accommodate interviewees who wished to discuss their experiences of and thoughts about 9/11 and subsequent developments at the local border even in the absence of any direct questions on these topics. As it turned out, everyone wanted to talk about these issues and, as a result, I was able to document and analyse how border residents were engaging with new top-down border securitization, filtered bordering, and Canadianness in the post-9/11 context.

In this chapter, interviewee and local press reports are used to convey something of the impact of 9/11 and post-9/11 border securitization as experienced by local border residents. As the interviewing and local press reporting spanned the 2001–4 period, it includes both more immediate as well as more retrospective reflections. The bottom-up perspectives on processes often associated with a re-bordering of North America after 9/11 offer an important supplement to the views of more powerful political and economic elites at both national and regional levels. I reserve a fuller discussion of both pre- and post-9/11 filtered bordering and Canadianness for subsequent chapters.

Experiencing 9/11

Interviewee accounts of 9/11 often included references to how news of the crashing of four hijacked commercial planes in New York, Washington, and Pennsylvania led to immediate concern for family, friends, and friends of friends living in the U.S. who might be directly affected. In addition to fears for those in the U.S., respondents also emphasized their own sense of more direct vulnerability to anti-American global terrorism as a result of their geographical proximity to the U.S.

Melissa, interviewed just three weeks after 9/11, described how an initial impulse to move to Toronto, just over an hour's drive away, was replaced by the idea of moving to a much more distant place: "Northern Saskatchewan [sounded] OK, you know ... like get away from this area, because we're ... so close to America." Jessie also recalled how "a lot of people didn't want to go out of their homes" and everyone was formulating plans in the event of "invasion" to "just go, drive north, as far as we can get away from the border." Looking back from the spring of 2004, when she was interviewed, Paula recalled thinking, "We're [Canadian Niagara] next [as a target for attack] ... just because we are so close [to the border]."

Several interviewees remembered fearing that a major hydroelectric generating station located on the Canadian side (but supplying the U.S. market) might be a target of terrorist attack. Rachel recounted that on the day of 9/11, people were saying, "It's the beginning of World War Three!" and that she thought, "[Because] we provide power for all of the Eastern Seaboard and New York and Washington ... [and because] it almost seems like we're a part of the States ... aggression would be taken out on us ... if there were more attacks to come."

Brenda interviewed just a month after 9/11, described a friend calling her on 9/11 to tell her to retrieve her child from school – an action supported by the school principal. She had first thought that this friend was "overreacting … because they're [terrorists] not going to be bombing an elementary school," but after her friend suggested that the nearby generating station might be targeted and that school children might be suffering "emotional danger," she followed the advice. Bill, also interviewed close to 9/11, recalled that on that day he had "the worst feeling [he'd] ever had. It was eerie," he said, "because normally there's a helicopter or a plane flying over [but] even the birds weren't flying that day … we were watching it on TV [and] I just couldn't believe what was happening."

He recalled heightened security at the power plant and rumours that Niagara Falls, Ontario, was one of three Canadian "vulnerable spots" targeted by global terrorists. Others described how local anxiety was exacerbated by workplace alerts and shutdowns, and Amanda recalled the "freaky" experience of being woken by the night-time sound of low-flying aircraft.

Those with relatives or friends employed at the border were anxious about their safety, their concerns increased by bomb threats that led to a temporary closure of the Queenston-Lewiston and Peace Bridges on 12 September 2001 ("Bomb Scares Shut Peace, Q-L Bridges: U.S. Customs Investigating Reports," *Niagara Falls Review*, 13 September 2001, A1). Bomb threats were not new – indeed, as one Canada Customs regional manager commented to the local press in 1999, "We get a number of them [bomb threats] each year" ("Bell Protestors Warned of Bomb Threat," *Niagara Falls Review*, 1 May 1999, A1) – but previous incidents had been less widely publicized and did not engender the same kind of fear.

In the immediate aftermath of 9/11, the Canada/U.S. border at Niagara continued to function, but in dramatically altered ways, as a U.S. level-one alert required U.S. Border Patrol and Customs agents to check every vehicle and person entering the country. While there had been previous periods of intensified local surveillance at the bridges, including after the December 1999 arrest of the so-called Millennium Bomber at the British Columbia–Washington border ("New Year Starts with Bridge Gridlock," *Niagara Falls Review*, 3 January 2000, A8), interviewees described the post-9/11 securitization as unprecedented.

Many interviewees shared how they had been particularly shocked at the closure of one of the Niagara bridges – the Whirlpool Rapids Bridge (referred to as the Lower Bridge by many Canadian border residents) – used by frequent border crossers enrolled in the fast-tracking preclearance CANPASS (Canadian Passenger Accelerated Service System) program.[1] The elimination of this crossing, combined with gridlock created by the bomb threats and intensified inspection at the bridges that remained open, heightened the sense of chaos for local residents with long histories of frequent border crossings.

The disruption to everyday life was immediate as people tried to return to their "home" side of the border. As Hannah recalled, "Everybody was rushing to get over [to the U.S.] … [if] they had family – or to come back [to Canada] … rushing to get back because they were closing all the borders." Tessa added her recollection of how "some people … got stuck," including the relatives of a close friend working on the U.S. side who "couldn't get back." Some found themselves offering psychological support to anxious American tourists trapped on the Canadian side.

While only the Whirlpool Rapids Bridge experienced more than a temporary closure, congestion produced effective immobility at other bridges. A September 12 headline declaring "Borders Open, Security Tight" (*Niagara Falls Review*, A1) was followed the next day by a story about the Peace Bridge, titled "Traffic Stalled at Beefed Up Border" (*Niagara Falls Review*, A2), and the public was asked to "postpone or delay any border crossing by vehicle … until long lineups can be cleared."[2]

In Niagara Falls, Ontario, and Fort Erie "trucker towns" were set up for drivers who waited to be directed in batches towards the bridges, while churches, the Red Cross, and other volunteers and businesses donated time and food to care for stranded travellers.[3] Interviewees emphasized the feeling of being unmoored by the new reality as they grappled with dramatically restricted cross-border mobility. Despite an early headline that claimed "Traffic Flow Back to Normal" (*Niagara Falls Review*, 22 September 2001, C9), this was not the local experience.

Brenda talked about seeing the closed Whirlpool Rapids Bridge and experiencing a sense of being in a "whole new era." The day of 9/11, she said, changed her perception of border life, because "until then it was just, you cross a bridge and you're in the States … It was no big deal." While she had previously thought about proximity to the U.S. in terms of the benefits for tourism and related employment, after 9/11 she felt newly endangered living "right beside the border," because "they [were] checking people" and there were reports that terrorists had "gotten across our border." While the latter claim turned out to be incorrect, Brenda was one of many who expressed a sense that Canadian Niagara was directly implicated and threatened by the new so-called War on Terror.[4]

In the first interview that I conducted after 9/11, dual citizen Dylan was angry and confused, telling me, "It's just hard for me to comprehend what is going on," and adding, "They [terrorists] definitely took my sense of security away." He shared his anxiety about the safety of local border bridges saying:

> What if I'm crossing this border one day and some person wants to blow up the bridge? ... There's absolutely nothing to stop you in Canada from crossing over to the States, until you get into the States. So how do I know that the person ... behind me doesn't have a bomb in their back seat and once they stop in the middle of the bridge, push the button, and I'm the car ahead of them?

This imagined scenario of global terrorists being present in the borderland itself was echoed in a November 2001 *Niagara Falls Review* editorial that urged locals crossing "to go to a Bills [football] game or do a little shopping" and to support the securitization project by being patient with the increased regulation of flows across Niagara border bridges. The editorial asked rhetorically whether readers would "really want to be stuck in the middle of the Rainbow Bridge, waiting to cross and wondering if the guy next to you is willing to die for a cause?" ("The Need For Security When Crossing The Bridge," *Niagara Falls Review*, 1 November 2001, A4).

While the alleged physical threat to border bridges and hydro infrastructure did not materialize, Paula recollected the new sense of danger associated with living at the border:

> Just living so close to it [the border] ... they could drop a bomb on us at any time. I think it's more real to us [at the border] ... You think you're immune ... to certain things like 9/11 and all the bombings and stuff like that, but in reality, we could be next ... when you see the planes flying around in the sky in circles ... it's like, "Are those good guys or are those bad guys?"

Along with personalized concerns for family and friends in the U.S., there were also more public expressions of cross-border solidarity.

Brenda noted that the local newspaper had "printed the ... American flag on one side, the Canadian flag on the other side," and even almost a month later everyone on her Canadian street had the American flag in their windows. "Until that point," she added, Canadian border residents had not "related ourselves to them [Americans] ... you would not see an American flag at all ... [but] now they [were] plastered everywhere." Jill, interviewed three weeks after 9/11, described how, like fellow Canadian border residents who were displaying American flags on cars, trucks and houses, her family had included the U.S. flags on their backpacks and uniforms because "we have American relatives, we lived in the States, we're very close. It hit us very, very hard. It hit us personally ... I'm proud to be Canadian but I'm also proud to have American friends."

Avery, also interviewed close to the events, commented that 9/11 had led to a "weird" and "kind of strange" shift from a previous everyday anti-Americanism, whereby Canadian border residents would say, "'Oh, the Americans, they're doing this' and you didn't really like them," to a situation where locals were flying American flags and "all of a sudden ... we're their best friends." Dual citizen Ashley shared that in this period her sister told their American mother that this was the first time she had felt "proud to be [part] American."

The proliferation of American flags was accompanied by municipal-level offers of cross-border emergency assistance and a vigil organized by Canadian-side civic leaders. A *Niagara Falls Review* report on the vigil noted that 1,500 people had attended and quoted a Niagara Falls, Ontario, alderman as saying, "We're a border town, we consider them [Americans] our neighbours, we work and breathe with them every day" ("Vigil Honours Neighbours," *Niagara Falls Review*, 19 September 2001, A1). The presence of an Imam from the Islamic Centre of Niagara Falls at the vigil, however, did not preclude intensified forms of internal exclusion, as the mosque in St. Catharines was targeted by arson and Muslim university students shared a new sense of vulnerability to everyday racist targeting that linked them to the terrorist attacks.[5] One of the interviewees, Ameena, who was of South Asian ancestry, reported that after 9/11 children "of different races" in Niagara Falls, Ontario, were being told by "white children" that "your people killed all the people in the Pentagon" and, even more bluntly, "your people killed our people." Narratives of 9/11, then, revealed shifts in interviewees' sense of border space and identity both in terms of the relationship to the other side of the river and, within Canadian Niagara, in terms of intensified forms of ethnoracialized Othering.

National media reporting in the immediate post-9/11 period often focused on the negative impact on commercial cross-border flows – most apparent at the Windsor-Detroit Ambassador Bridge, where traffic was backed up for 36 kilometres – and local automotive plants, which, dependent on cross-border transportation, closed down (Andreas 2005, 457). Canadian government and industry concern over combining the new U.S.-led securitization with commercial cross-border flows precipitated the signing of a new joint Canada-U.S. Smart Border Declaration and Action Plan on 12 December 2001 promising more action than border-related agreements of the 1990s (Pratt 2005; Clarkson 2006, 602–3).[6]

In Canadian Niagara, local business leaders on the Canadian side were offered early reassurance by the Ontario provincial finance minister – in a talk to the Chamber of Commerce in Fort Erie less than two weeks after the 9/11 attacks – that the "crisis," while marked by apparent re-bordering, might lead to "technological advancements" in security that would eventually speed up, rather than slow down, border crossings ("Terrorism Has 'Huge Impact' On Tourism, Flaherty Told," *Niagara Falls Review*, 22 September 2001, A3).

Some Niagara elites, in turn, attempted to link long-standing visions of official cross-border regionalism to the new securitization by portraying these as congruent with the new Smart Border project. The Canadian federal government's December 2001 budget, which directed nearly $8 billion to anti-terrorist and border security–related projects over five years, was described in a *Niagara Falls Review* editorial, for example, as "very good news for Niagara," as both Niagara Falls and Fort Erie, it suggested, would "cash in" on federal investment in border infrastructure ("Judge Martin's Budget on Its Impact on Borders," *Niagara Falls Review*, 11 December 2001, A4).[7]

Subsequent investment in local Canadian-side border infrastructure would include changes to the bridges. In the case of the Peace Bridge, this included an expansion and reorganization of the bridge plaza and related buildings, feeder roads, and highways as well as the relocation of some border functions away from the bridge itself, for instance, the placing of a preprocessing centre on the outskirts of Fort Erie in 2004. Along with investment in the physical infrastructure there were new processes of border inspection and surveillance that involved increased staffing, new technologies and procedures including increased binational coordination and information sharing.[8]

Much of the more applied scholarship on the post-9/11 Canada/U.S. border is preoccupied – as are government and industry interests – with how to combine post-9/11 border securitization with the expansion and speeding up of cross-border commercial traffic. Other more critical scholarship focuses on how processes of post-9/11 securitization intensified ethnoracialized filtering (e.g., Pratt and Thompson 2008; Pratt 2010; Mountz 2011). My study had not anticipated a focus on intensified securitization, but the interviews and local press allowed me to document how this was experienced, imagined, and engaged in by border residents of Canadian Niagara. In what follows, I focus on two of the recurring topics that emerged from this material, notably the link between securitization and slowed and reduced cross-border mobilities as well as more onerous border inspection processes. Following discussion of both of these, I examine the ways in which border residents offered both critique and legitimation of these developments.

Border Im/mobilities

The most significant impact of post-9/11 securitization mentioned by interviewees was that of slowed cross-border mobilities and congestion in the border transportation corridors. As "mobilities" scholarship reveals, transportation is connected with "complex patterns of social experience" (Sheller and Urry 2006, 208) and the "friction" of traffic congestion more specifically, can "interrupt the prevailing narratives of unimpeded flows of goods, ideas, money and people accompanying market led globalization" (Truitt 2008, 4). These insights are relevant for understanding interviewee descriptions of how their everyday mobilities were disrupted by top-down securitization, which, as Bill noted a couple of weeks after 9/11, had made the local border "so tight it's not funny." The new reality of slowed cross-border trips and congestion contrasted with what had been a period of many local initiatives explicitly aimed at both speeding up and expanding commercial flows through Niagara. As Judith noted, there had been the construction of a newer "Lewiston Bridge to take the trucks," new booths at both the Rainbow and Peace Bridges, and a pre-clearance CANPASS program at the Whirlpool Bridge that allowed enrollees such as her parents to "go through the preliminaries away from the border." The events of 9/11 were experienced as disrupting the trajectory of these local developments.

Following the initial chaos of 9/11, there were recurrent periods of traffic gridlock. Between March 2002 and January 2004, for example, the new U.S. Homeland Security colour-coded threat assessment system was raised to orange (the second-highest level) five times and each time the result was a traffic "tail" of trucks backed up along commercial transportation corridors that were simultaneously everyday thoroughfares for local residents ("Bridge Officials Optimistic Lower Alert Will Ease Traffic," *Niagara Falls Review*, 10 January 2004, A1). Jim from Fort Erie complained about the experience of driving home from university and finding stalled trucks blocking his highway exit, saying, "You can't get home ... it takes forever, and nobody wants to play with those trucks, that's nerve-wracking." Coleman's (2005, 202) reference to "local collisions" as a result of "incoherent and countervailing" centrally driven border projects is relevant to the tragedies that unfolded in Canadian Niagara as stalled commercial traffic led to four highway fatalities (and other injuries) within a six-month period in early 2003 ("Despite Signage, Bridge Accidents Continue: Latest Victim Remains in Critical Condition," *Niagara Falls Review*, 25 July 2003, A3). Congestion intensified pre-existing local concerns about the negative impact of commercial traffic on border residents' health. *Niagara Falls Review* reporting revealed pre-9/11 concern on the Canadian side about air and noise pollution associated with increased flows, and on the U.S. side there was a history of resident mobilization against new or expanded border bridges for this reason.

After 9/11, the topic of air and noise pollution emerged again in *Niagara Falls Review* letters to the editor from Canadian-side residents debating an ultimately unsuccessful proposal to introduce a truck crossing at the railway bridge in Niagara Falls.[9] Studies showing the negative impact of idling truck emissions on resident health in both Cuidad Juarez, Mexico, and Buffalo, New York (see Lwebuga-Mukasa et al. 2004), were also raised in the context of debate about a planned expansion of the Peace Bridge. On the U.S. side, for example, the Buffalo West Side Environmental Defense Fund, a citizens group, argued that a new bridge should be located further away from major population centres and that there was a need to "compensate Buffalo for the unjust burdens and costs incurred as a result of trucks passing through."[10] On the Canadian side, a Fort Erie councillor noted that the "sporadic" local expressions of concern about both air and noise pollution of the previous decade were now intensifying, due to U.S.-bound trucks being "backed up for more than 10 kilometres," and the town council began preparations for air-quality testing ("Fort Erie Wants Toxic Diesel Fumes, Effects Monitored," *Niagara Falls Review*, 24 February 2004, A1).

While some interviewees referenced how delays were affecting the border-related livelihoods of family and friends (e.g., those in the trucking industry), most focused on the disruption of their own patterns of frequent border crossing. Even when the initial chaos eased, some had decided not to resume their cross border-activities. Jessie, for example, described how she did not cross for over a year for fear that the bridges might become terrorist targets. Kerrie too recalled, "We didn't cross for a while. Nobody really wanted to have anything to do with crossing."

Fears about the vulnerability of the bridges to terrorist attack were, over time, replaced by frustration with the inconvenience of long lines that discouraged discretionary crossings. Highway warning signs stating "Bridge full, four hour wait" were recalled by Zak, while Kevin described traffic on the Peace Bridge "backed up all the way onto the Canadian side … [as U.S. authorities] were checking every single vehicle and really going over them with a fine-toothed comb." Dave, interviewed in 2004, stated, "We [still] don't go over the border unless it's absolutely necessary" and several others in this period still complained about the inconvenience of unpredictable crossing times. Some mentioned, for example, the challenge of timing cross-border trips to attend major league sports events in Buffalo, and Amanda related how a cross-border family reunion was derailed by traffic congestion that led relatives to turn around before even reaching the border crossing.

Regional leaders, meanwhile, portrayed Smart Border initiatives as the solution to blocked or slowed local mobilities by linking local concerns to the broader project of facilitating secured continental commercial flows. The four highway fatalities in early 2003, for example, led the general manager of the Niagara Falls Bridge Commission to emphasize the need for an expanded deck, and new express and truck queue lanes at the Queenston-Lewiston Bridge (see "Despite Signage, Bridge Accidents Continue: Latest Victim Remains in Critical Condition," *Niagara Falls Review*, 25 July 2003, A3). Bridge authorities also acknowledged local anxiety about truck emissions, but reiterated that the solution was not to reduce crossings or move bridges away from more populated areas but rather to speed up traffic flow.

By 2003, one officially proffered "solution" to slowed cross-border mobilities was enrolment in a new, binational Nexus pre-clearance program designed to fast-track the border crossings of pre-approved, so-called low-risk travellers over the Peace and Rainbow Bridges. The Nexus program has been described as having emerged "as a context contingent response to the contradictory imperatives of national securitization and economic facilitation" (Sparke 2006, 153). In fact a Canadian pre-clearance program called CANPASS had already been introduced to Canadian Niagara when access to the Whirlpool Bridge became restricted to prescreened travellers in 1998. The conversion of the bridge to a CANPASS-only crossing was intended to offer a "fast-track" option for frequent crossers, but it was this bridge that was completely shut down on 9/11. In 2002, it reopened but then closed again, finally reopening as a Nexus-only crossing in the spring of 2004 and joining the Nexus-only lanes that had already been introduced at the Peace and Rainbow Bridges in 2003 ("Canpass Back at Whirlpool," *Niagara Falls Review*, 27 April 2002, A2). The general manager of the Niagara Falls Bridge Commission pitched Nexus enrolment to Niagara residents by portraying this as the solution to their experiences of cross-border congestion – even while suggesting that such congestion was a "myth" ("Border Myths Addressed by Bridge Official," *Niagara Falls Review*, 10 April 2003, A3). When long lines formed at the Queenston-Lewiston and Peace Bridges due to a U.S. orange alert over the holiday period of December 2003, he again took the opportunity to remind residents that enrolling in Nexus was "the best way to get over the border quickly" ("Bridge Officials Optimistic Lower Alert Will Ease Traffic," *Niagara Falls Review*, 10 January 2004, A1).

Initially, however, the binational Nexus program was not as popular as the phased-out CANPASS program had been. CANPASS had offered rapid re-entry to Canada for free and by 1999 had close to 40,000 users (both U.S. and Canadian) in the region ("CanPass at Peace Bridge Aug. 16," *Niagara Falls Review*, 12 August 1999, A1). Fewer border residents, however, appeared willing to enrol in the Nexus program, which involved a more complex process of one-on-one security interviews with both Canadian and American authorities, provision of a photo and biometric data (fingerprint and retina scanning), and a significant fee that was initially eighty dollars but was later reduced ("Binational Tourism Alliance looks to Boost Profile of Nexus Program," *Niagara This Week*, 18 April 2007, 9).

Jim was explicit about the negative impact of what he perceived as U.S. top-down securitization on Canada/U.S. bilateralism as well as the everyday relations between "border towns," suggesting that use of biometrics, in particular, represented an affront to border residents:

> I think that [President] Bush wanted an iris scan, and that's going a little far. That definitely damages cordiality between border towns and countries, because it's ... [a] lack of trust. I mean ... maybe it's not as [much of] a personal thing as ... trying to keep away the terrorists, but you know, it definitely doesn't make anyone want to cross the border on a frequent basis.

This comment echoes elite explanations for the initially low rates of overall Nexus enrolment (not just in Niagara) as being due to "perceptions of the cost of privacy invasion" as well as a "lack of infrastructure" (Salter and Piche 2011, 944–5). Overt resistance was apparent in Canadian Niagara when the 2004 conversion of the Whirlpool Rapids Bridge to Nexus-only status led some small businesspeople on the Canadian side to complain that the resulting reduction in cross-border traffic threatened their livelihoods. One businessman suggested that there had been a reduction from 800 crossings a day under CANPASS to only 150 crossings a day under Nexus ("Bridge Changes Still Impacting Businesses," *Niagara Falls Review*, 13 April 2004, A3), and a letter to the editor critiqued the new security fencing that made the Whirlpool Bridge "look like Guantanamo Bay" rather than a welcoming "international gateway" to the city ("Downtown Merchants Are Frustrated," *Niagara Falls Review*, 6 April 2004, A4). There were calls for the bridge to be returned to its pre-CANPASS status, when it was open to everyone and, as a one barber claimed nostalgically, "people used to walk across [the bridge] to get haircuts" ("Whirlpool Bridge Reopens: But Merchants Say Nexus May Hamper Business," *Niagara Falls Review*, 1 March 2004, A3).

While critics suggested that new border investments were hindering rather than facilitating cross-border flows, others were more positive about their potential. One letter writer urged readers to recognize Nexus as "one of the best things to happen in cross border travel" and to stop "thinking of the cost" ("Leadership Blamed for Bridge Woes," *Niagara Falls Review*, 14 April 2004, A4). Others however, described the program as accessible only to the "privileged" who could afford the enrolment fee. As the latter comment suggests, the expansion of expedited cross-border mobilities through Nexus was received in class-differentiated ways at the local level.

The issue of classed differentiation may offer an explanation for the fact that while interviewees expressed their frustration with slowed and, as I discuss below, more intensely surveilled cross-border mobilities post-9/11, they did not discuss enrolment in Nexus. Although Dan stated that he "would like access to the States whenever [he wanted] without a hassle and not having to wait in a line" and Zak wanted to see a "Fortress North America with Mexico included, and ... a super pass that would get you through," for example, neither of them mentioned pursuing the Nexus option. That there were financial and other barriers to enrolment among a population with a strong interest in cross-border mobility supports Sparke's (2006) argument that Nexus can be understood as contributing to neoliberalized citizenships marked by "new transnational mobility rights for some and new exclusions for others" (153). The introduction of Nexus to Canadian Niagara may have simply deepened existing class-differentiated cross-border mobilities by speeding up the crossings of the economically privileged.

Another factor influencing the slow uptake of Nexus when it was first introduced for lanes on the Peace and Rainbow Bridges in 2003 was that the exchange rates were less favourable for Canadians. By 2004 however, a stronger Canadian dollar was leading at least some to resume limited cross-border shopping for economic reasons. Zak for example, described how his grandparents, having calculated exchange rates, U.S. road and bridge tolls, the hassle of border inspections and waiting times, concluded that they would save " about four cents a litre" on gas and so had begun crossing the border once again. He added, however, that these trips made economic sense only because they "have all the time in the world, and they count every penny." These seniors who had to "count every penny" were apparently willing to endure a slower and more surveilled border but interestingly did not, it seems, enrol in Nexus.

Border Inspection

Closely related to the topic of how new forms of securitization produced blocked or slowed cross-border mobilities in the post-9/11 period was discussion of what were perceived to be intensified border inspections. As the previous chapter made clear, learning how to navigate border inspections was part of growing up in the border region, but the rapid changes that followed the events of 9/11 made people less confident in their ability to do this successfully. At the same time, however, many border residents continued to benefit from privileged insider knowledge of the changes, due to their personal connections to those working in the border infrastructure. Dale, for example, felt that he was kept abreast of security developments through border workers whom he had known since childhood, especially one who "hated the border" and "disagreed with ... most of the rules" and as a result was particularly forthcoming.

The descriptions of post-9/11 changes in border inspections often focused on and critiqued new documentary requirements. Jim was one of those who reflected on the contrast with the pre-9/11 period when he crossed with "nothing [no identification] on [him] ... [but now] everybody has to show ID and now they question everybody in the car ... not just the driver, but every single person ... what your citizenship is, and where you're from."

In a description of how her first trip across the border after 9/11 took an unexpectedly long forty-five minutes (despite being the only car going through), Barb noted that this was because border officials were now "probing into everything and they want to see ID." Paula also recounted how a cross-border trip was unexpectedly thwarted when a U.S. official rejected her sister's attempt to use a student card as identification, an experience that led her to avoid any further attempts at crossing. Tim, whose father worked in the trucking industry, discussed how much stricter the border crossing had become now that that they "ask more questions [and] ... want to see your ID almost 90 per cent of the time."

The new requirements were of concern to Dylan, a dual citizen interviewed right after 9/11, who noted that he now needed "citizenship and papers and passports" and was anxious about the need to keep such documents in the car, where they might be stolen. As he described it:

> [I'm] getting nervous because I don't know how I'm going to reply [regarding border officials' query as to citizenship] ... If I say "Canadian" I'll just have to show my ... Ontario [driver's] licence right? ... But if I say "American," then I'm going to have to bring all this paperwork with me and "Here's blah blah blah and according to the consulate of you know."

That new documentary provisions were also deepening existing citizenship-based differentiations between border residents was indicated by a *Niagara Falls Review* report citing the concerns of the executive director of the Niagara Folk Arts Multicultural Centre (of Niagara, Ontario) about a new U.S. proposal to require visas of landed immigrants in Canada ("Immigrant Workers Condemn Latest U.S. Policy," *Niagara Falls Review*, 5 November 2002, A3). Canadian-side changes that same year requiring landed immigrants to acquire and present new Permanent Resident Cards to re-enter Canada by commercial carrier were not highlighted but had a similar effect (Browne 2005).

In addition to new processes surrounding the documents that mediate the "traveler's relationship to forms of state power" (Greenberg 2011, 90), Canadian border residents focused on what they experienced as newly intrusive questioning by U.S. officials. The latter appeared to increase their sense of the border as a "permanent 'state of exception'" where "one may claim no rights but is still subject to the law" and subject to increased "confession of all manner of bodily, economic, and social information" (Salter 2006, 169–70). Kerrie, for example, noted that border officials were now asking, "Whose car I was driving? Where [did] I go to school? How [did] I pay for my car? … Where I was going? How I knew these people? Their address? If they lived with their parents? What their parents did? What my parents [did]? … I don't know that it matters how I [paid] for my car!"

Rachel found on post-9/11 visits to U.S.-based friends that U.S. border officials checked the entire car and asked questions like "Where are you going? Why are you going there? … How did [you] meet them? How long have [you] known them? Who are they? Their names? What they do? While [you're] there, what [are you] going to be doing?"

Miles, whose father was a border worker, noted in his 2004 interview, "In the last couple of years, I've gone to a couple of concerts [in the U.S.], and they [U.S. border agents] really kind of treated you like a criminal." While conceding that some crossers needed to be questioned in greater depth, he didn't "understand why they would ask us, because we didn't really look like the criminal type, it's just a family of four going over the border." The sense of being newly criminalized was acute for the largely white interviewees of my study who, as previously discussed, had experienced relatively eased cross-border mobilities as children and teens. Dan, who had crossed regularly (to attend Nascar races), noted in 2004 that "99.99 per cent of [border resident] Canadians that cross into the States are meaning no harm, it's part of their daily routine," but that, post-9/11, the U.S. border authorities were really trying to "throw you off" by asking questions they never asked before. The new inspection process, he added, made him "kind of angry … I feel like, 'Hey, I grew up here and I've come across your border and I know I'm not meaning any harm and I'm going to give your economy money [through cross-border shopping and other activities] and you're giving me such a hard time to get across.'"

Interviewed a month after 9/11, Lisa, who as a young adult had worked as a janitor on one of the bridges and crossed frequently as a younger person, reported that she had not yet resumed cross-border movement, because she was "afraid" that it was "very strict" and she expressed particular concern that Canadians returning with undeclared goods might now be "caught."

The changes unsettled these border residents' sense of familiarity with the workings of the local border, and many were frustrated that post-9/11 bordering was not "smart" enough to recognize them as self-evidently safe and desirable local border residents who had crossed frequently in the past. Interviewees' emotional accounts of their experiences of post-9/11 border inspections reveal how the "systems and symbols of state authority that constitute sovereign statehood are as much structures of feeling as they are structures of force: congeries of affect as much as of action" (Chalfin 2008, 522). Accounts of a local border that was experienced as less predictable in terms of the length of time required to cross as well as more intrusive in terms of border inspection reveal the embodied experiences of a shifting "emotional geography" (Sheller and Urry 2006, 216) for these Canadian border residents.

Critique and Legitimation

Given my interest in how Canadian border residents not only experienced and perceived top-down border projects but also engaged with such projects, I was attentive to indications of general support, acquiescence, or resistance to post-9/11 securitization at the local border. The opportunity to hear from those living close to a border that was changing in response to an alleged global terrorist threat was revelatory of the "experiences of ordinary people encountering the fears and insecurities that are both cause and consequences of the deployment of security-and-surveillance apparatus" (Lyon and Murakami Wood 2012, 326).

As mentioned, for many Canadian Niagara border residents, the new securitization was understood in personal terms due to intimate connections to Canadian border workers. Such connections meant that many also understood that there was a lack of full congruence between border workers and state border projects (as revealed, for example, through the tacit facilitation of minor smuggling by locals). The disjuncture between border workers and the Canadian state was also made salient in the context of recurrent border-worker protests and stoppages that signalled discontent with various aspects of their state-determined working conditions.

I was interested, then, in how some of the interviewees who offered expressions of support for the role of Canadian border officials in post-9/11 securitization also criticized what they perceived to be their newly expanded regulatory power and responsibility. Some of this critique centred on perceptions of a mismatch between such power and the phenomenon of younger Canadian university students being hired as seasonal or part-time border officials (as some of the interviewees themselves had been).

Kerrie, for example, shared how she used to think that customs jobs were "really hard to get" and required a university degree, because the workers "seemed so much higher than everybody else and had so much more control and power," but she had since learned that "all you have to do is apply online and go write a test." Having done this herself, she had learned that only a high school diploma was needed and that getting an interview was "pretty easy," which she thought was "scary because the kids I know that work there that are in first year [university] now ... [and] I don't think they should be protecting Canada."

Dale also expressed some discomfort with the tendency of young "twenty-somethings" to engage in "power trips" in the course of border employment. He referenced a border-worker acquaintance whom he described as using discretionary power to harass American crossers, for example, by asking U.S. travellers to put out their cigarettes and then pulling them over to secondary inspection if they resisted.

This kind of age-related critique could be extended to the U.S. side as well. Sheila (whose mother had dual Canadian/U.S. citizenship) complained about what she considered to have been overly harsh treatment of her father at the U.S. border by a young U.S. official who had told her father that his driver's licence no longer provided sufficient documentation for entering the U.S. She was frustrated that her father had been treated discourteously by this (in her view inappropriately empowered) "kid" of nineteen:

> [He] was just giving it to my dad ... [saying], "You need to have your birth certificate. If you don't ... how do I know it's you? ... I can make you turn around and go all the way back and get it. I can pull you into

> an inspection." And my dad's like, "I'm sorry, every time I've come over here you've never asked for it," [and] ... my dad wasn't rude at all, he ... just kept his cool ... but ... I [was thinking] ... "You're younger than I am. Don't talk to him that way"... [and] to have him do that to my father ... turned me off ... [going] over there [to the U.S.].

She described further how, despite having a friend who loved being a customs officer and encouragement from her own parents to apply for what they considered this "good job," she had abandoned an application for customs work in favour of continued employment at Walmart. The decision was due to personal discomfort with the structural power involved. As she explained, "I can't do that to people. I can't make them feel worthless. I can't sit there and go 'What did you buy? Citizenship? What did you do?' That's not me ... I'd be like 'OK, go ahead.'"

As mentioned, such direct or indirect critiques of border work and workers, however, coexisted with more supportive portrayals. In referring to a work-to-rule "strike" by immigration workers for example, Lisa suggested that the resulting traffic congestion only made it clear just "how much stuff they have to do ... [and that] you need them there." She went on to describe non-terrorist examples of important border "protection," including against "people who bring guns over," the "drug industries," and child abductors, because "if you bring over your kid's friend, you have to actually have a note, and they'll make the kid call the parents ... They do that all the time." Anne, from a lower-middle-class family in Niagara-on-the-Lake, was sympathetic to border workers stating that their surveilling behaviour "was part of their job." More difficult exchanges with officials, she felt, could easily be avoided if crossers would avoid "antagonizing" them, because "there's no need to have a situation if you haven't done anything wrong."

Despite frustration with some of their own border experiences, most interviewees echoed the dominant discourse in suggesting that, despite the inconveniences for them as border residents, increased border securitization was a necessary part of the newly localized "War on Terror." Miles, who complained of being "criminalized" by a U.S. border inspection process that was "a little overboard," nevertheless concluded, "I guess they've got to do it." Mark from Niagara Falls, Ontario, described the changes being experienced by border residents as acceptable, because border officials "have to check security and do what they can." Greater border security was needed, stated Peter, because "you don't know who is crossing now," and as Anne noted, she now realized that "not everyone in the world had good intentions." That securitization at Niagara was part of a larger anti-global terrorism effort was suggested by Jim, who while initially claiming that the Americans had "gone too far" in their actions at the border, followed this by saying, "Then again, you never know when it is in your own neighbourhood, because there were some members of Al Qaeda in Buffalo ... plus the bomb threats on the bridge, and [while] obviously none of them have come to pass but ... you just can't take that kind of a risk."

Some even used their local knowledge and observations to criticize U.S. border workers for what they saw as insufficiently rigorous security post-9/11. Dave, for example, expressed scepticism about U.S. efforts, recalling, "We were waiting forever [at the Queenston-Lewiston Bridge], and we get to the front of the line, and it's just one guy sitting there watching all the cars go through this narrow opening. And that was their terrorist protection!"

He went on to claim, "They say they're [U.S. border officials] cracking down ... [but] I haven't noticed much of a change." Changes at the border, he suggested, were creating more hassle than real security. Yvonne critiqued her own still facilitated crossing – which she suggested was linked to a problematically gendered filtering that reduced surveillance of young women, commenting critically that she "flew right by [the border] ... and they didn't even care, because there was a few girls in my car, and we just looked harmless ... There should have been more security."

Referring back to the pre-9/11 period, Lisa shared her newer view that the Canadian customs officers of that time had been too lenient because she had herself "hung out in the customs officer's booth where people [were] going through and ... it got to the point where they were just going, 'Okay what's your citizenship? Bring anything over? No? Okay, go through.' ... Some of them don't really care."

Similarly, Jill shared her shifting views, noting that a couple of weeks before 9/11 she had been saying to her husband, "'Why do we still have this [border]? It's ridiculous' ...[because] you'd see a car full of young guys and think, 'Well, they should be pulling them over,' but they would just go straight through. I just found it very lax." While she had thought previously that the pre-9/11 border didn't "make a lot of sense" (even while suggesting that "young guys" should be "pulled over"), following 9/11, she claimed that she was looking at it "a different way ... [because] maybe we do need that border to filter out people."

Chuck also agreed with the increased security post-9/11 despite the fact that this meant more thorough inspections for him as a frequent crosser. The day before the interview, he had crossed to attend a Super Bowl party at his best friend's house in the U.S. and found that he was now being inspected upon entry and exit and that the process was more in depth because "they ... ask for identification ... ask me a few more questions ... ask me to pop my trunk but ... I can respect [this] due to the circumstances ... I find it acceptable." He added that such measures would negatively "affect [cross-border] relations and ... the tourist industry between the border towns" but suggested, "It's what has to be done in today's society." His view that the Canadian side needed "to be tightened" represented, he acknowledged, a dramatic shift from his pre-9/11 views that "the border was more symbolic than anything else."

Several others also shared how 9/11 had led them to revisit previous views of state bordering practices. Brenda, in an interview just after 9/11, described how she was surprised and concerned when she was younger to learn that border guards had the power to strip-search crossers, but added that the events of 9/11 had led her to rethink this. Now, she said, she felt "it's good if they [border workers] pull me over, because that means they're doing their job." Tim, who described an unprecedented experience of having his car door "ripped apart" post-9/11, also avoided overt criticism of the border officials responsible, saying, "I don't know... I understand them doing their job and ... being careful," suggesting that as student crossers, he and his friend probably warranted additional attention. When pressed on this latter point, he only reluctantly acknowledged that there might be "stereotyping" (of students) by officials who "probably just assume we're troublemakers or something," but added the disclaimer "I don't know what to say," indicating perhaps some uncertainty about expressing critical views of state re-bordering in the post-9/11 context.

Conclusion

The unexpected events of 9/11 had a profound impact on Canadian Niagara and, therefore, on the unfolding of this project. Because I had already begun interviewing Canadian border residents about everyday life in the border region, I was able to capture something of how 9/11 and early post-9/11 border securitization was experienced, imagined, and engaged with by my respondents. The resulting material offers more grounded accounts of "re-bordering" linked to post-9/11 securitization, which can supplement those of national and regional elites whose views dominated the airways at the time and have since attracted the bulk of scholarly attention. The interviews and the local press reporting allowed me to document and begin to analyse the emotional and logistical impact of 9/11 itself on border residents. These sources revealed local understandings of the border region as newly vulnerable to global terrorism and the lived effects of an effective border closure that blocked familiar cross-border flows. The material presented also illuminates post-9/11 developments and, in particular, residents' experiences of slowed cross-border mobilities and what were described as newly onerous border inspections that altered crossings and unsettled local knowledge and confidence regarding border navigation. While generally supportive of the anti-terrorism rationale for border securitization, some interviewees drew upon personal links to border workers and close observation of new forms of border securitization to express scepticism about the efficacy of the new re-bordering. Their muted critique, however, suggests that in Niagara as elsewhere, the discourses of securitization facilitated state use of "extraordinary powers and the subsequent removal of policy issues from public debate" (Salter 2007, 315). In the following chapter, I look more closely at the nuances of this, as I delve more deeply into how border residents experienced, imagined, and engaged with filtered bordering in particular.

Chapter Four

Filtered Bordering and Borderline Lives

Introduction

Critical scholarship in both border and Canadian studies has begun to document how territorial and non-territorial borders, whether global or Canadian, serve less as walls or gates than as sieves that filter flows in differentiated and unequal ways. In this chapter, I look more carefully at interviewee accounts of bordering to better illuminate this phenomenon at Niagara. I then consider the implications of such filtering for border identities and "border culture" in Canadian Niagara.

In the earlier discussion of childhood and youthful border crossings in chapter 2, it was clear that as young people the interviewees were often aware of patterns of filtered bordering at Niagara and that part of growing up in the border region involved learning about the importance of performing "safe citizenship" at the border inspection. This and chapter 3, on the immediate post-9/11 period, begin to reveal how local residents experienced, understood, and engaged with unequally facilitated or troubled crossings that they linked to age and life course as well as intersecting classed and gendered positionings. Here I look more carefully at residents' accounts of filtered bordering, as I return to these earlier discussions and augment them with additional interviewee material related to bordering and Indigenous, ethnoracial, and citizenship positionings before and after 9/11. While I employ a pre-9/11–post-9/11 division in presenting this material, it is clear that there were significant continuities across this chronological divide.

Interviewee accounts of filtered bordering drew from personal experience with, and close observation of, the local border. These accounts, I argue, challenge both elite and everyday claims of a benign and even friendly Canada/U.S. border (see also Pottie-Sherman and Wilkes 2015). The borderline that emerges here is much closer to Bhandar's (2008) description of a boundary that is "constituted differently depending on the racialized, classed, and gendered bodies that pass through its mechanisms" (292). The narratives from Canadian border residents are revelatory insofar as they add to understandings of the dynamics of Canada/U.S. bordering. They also raise questions about border identity and culture in a context marked by the apparent intensification of "new hierarchies of value built around mobility" (Greenberg 2011, 91).

Filtered Bordering Pre-9/11

While my interview pool was not very diverse, the accounts of interviewees made it clear that both pre- and post-9/11 bordering was understood as differentiated and unequal, in the sense that distinct categories of border crossers (including, as discussed, children officially constructed as being "at risk" and youth as "risky"), encountered different degrees of scrutiny and surveillance or processes of "social sorting" (Lyon 2003) by border officials, which, in turn, facilitated, slowed, or impeded cross-border mobilities. One of the categories that emerged as significant in interview and local press reports was that of Indigeneity. Erin, who identified some of her relatives as Six Nations, for example, recalled that when she crossed with these family members, they would "just say 'Six Nations' and then they'd let you over." Others, however, linked Indigeneity to more difficult crossings. Dale recalled witnessing an incident where customs officers accosted "an Aboriginal woman in a car … [and were] wrestling with her to take the keys out of the ignition … saying that she had cigarettes" and remembered "sitting there in [the] car … thinking, 'Whoa, someone did something wrong.' I mean, she was screaming." While Dale was ambiguous about who committed the "wrong" that was witnessed, the story linked Indigeneity to a more fraught border crossing.

Mary, who worked as a customs officer in the 1960s, recalled that Canadian border officers' questioning of Indigenous crossers about their place of birth often heightened tensions. The annual border-crossing parade, an event that asserted 1794 Jay Treaty rights to free movement and exemption from duty mentioned in chapter 1, was described by Mary as the "Indian Day Parade" and she recalled that the Indigenous participants would "dress up" in regalia and that it looked like "a Western." She also described how female officers such as herself were kept away from the "Indian" buses whose passengers, she recounted, were described by other border guards as "drunk and hostile," a portrayal that she considered to be stereotyped and inaccurate.

The local press also included reporting on recurrent bridge-sited Indigenous protests. A *Niagara Falls Review* report on a 1999 Peace Bridge protest by those identified in the news story as part of the North American Native Warriors Association and allies, quoted a spokesperson asking cross-border travellers delayed by the political action to "understand that our rights have been violated for over 507 years" ("Native Protest Closes Bridge," *Niagara Falls Review*, 29 November 1999, A1). The following spring, when the same groups were reported in the local press as protesting "the treatment of aboriginal peoples by the governments of both Canada and the United States" other crossers experiencing the resulting delays however, reportedly "blasted their horns and shouted obscenities" ("Upset Motorists Honk Back at Blockading Natives: Peace Bridge Traffic Snarled by Protest," *Niagara Falls Review*, 7 February 2000, A6).[1]

Along with these references to Indigeneity and bordering, interviewees alluded directly or indirectly to many other forms of filtered bordering. Classed positionings, already discussed in relation to cross-border shopping, emerged in several other contexts, including discussion about wealthy U.S. vacationers crossing the border to visit summer cottages in Fort Erie. Mary, in further recollection of her employment at the border, described how fellow "nationalistic" Canadian customs officers felt that these seasonal visitors "should be shopping for their groceries and their liquor in Canada" instead of the U.S. and used their discretion to target these well-off border crossers for searches. In contrast, she discussed her own reluctance to search "working-class" Canadians returning from the U.S. with "back-to-school clothes for their kids," because she didn't feel that "it was [her] job to … lay into them." Her own in-laws, she recalled, would re-enter Canada "sweating" under three layers of "T-shirts … underwear, and socks," and so as a border worker, she chose to turn a blind eye to other "little roly-poly kids." While framed in terms of nationality, her distinction between wealthy U.S. cottagers and working-class Canadian cross-border shoppers pointed to class-based filtering, as did her nationalized but also class-infused morality of economic justice.

Mary's description of harassed wealthy cottagers, however, contrasted with the more common suggestion by interviewees that it was markers of poverty or subcultural marginality – not wealth – that attracted greater official scrutiny. Mary herself reported that she knew from personal experience that crossing the border in a "rusted car" could lead to greater hassle, while Zak suggested that his buddies attracted more attention at the border because they weren't "the cleanest-shaven guys" and might be driving "a beat-up car or a big pickup truck."

Barb contrasted childhood crossings with her mother and father, suggesting that the latter's "rough exterior" consistently produced more sustained questioning:

> It was always relatively easy to go across with my mother ... we never had any hold ups. They'd just basically ask where we were going, how long we were going to be there, when we [would] be returning, and that was about it. When I crossed with my father, on the other hand – he had a very rough exterior at that point in time ... the long shaggy hair ... into the late 1980s he was still dressing as a hippy – it just seemed to take forever to get across the border.

Several others also pointed to how lower-class markers could combine with other officially perceived markers of criminality in the pre-9/11 period to increase the chances of men with beards being "pulled over." (Ashley thought this was because facial hair was associated by border officials with "smuggling drugs.") Perceived markers of homosexuality were also linked to greater hassle; Nancy described a gay colleague adopting a strategy of removing his earrings before crossing the local border to avoid being targeted for greater scrutiny by border guards.[2]

As the previous discussion and the above examples suggest, gender played a part in filtered bordering, and many interviewees thought that males – and groups of young males in particular – were likely to be more intensely surveilled. Tom stated that an acquaintance who worked at customs had confirmed his sense that "if it's four guys sitting in a car," the border official would "usually ask more questions" and check "to see if they're high [on drugs] or if there's some sort of influence." Zak felt that the presence of young women eased the passage of young males across the border, because the latter were seen by border officials as "more innocent" and less likely to be "coming over to start fights." Tim, a lower-middle-class interviewee from Niagara-on-the-Lake, found that when he crossed with his girlfriend instead of his male friends he had "no problems." He also suggested, however, that young women crossing with young men were more likely to face more questioning (than if they crossed with females only), because a protective border official might ask them, "Do you know these fellas?"

Brenda also suggested that gendered bordering affected young women in particular ways when she described a younger female relative who was refused U.S. entry when she tried to go cross-border drinking "dressed in a little leather skirt." This young woman "came home absolutely flabbergasted and shocked that they [U.S. border officials] would not let her cross because they thought she was a prostitute."[3] Other female interviewees also expressed concern about the possibility of strip-searches at the border – something not mentioned by male respondents.

Kerrie, who reported that her female friends were often "afraid" of male officers, emphasized that she, in contrast to them, knew how to handle what she portrayed as highly gendered interactions with flirtatious male border guards. One time, when she was crossing with a female friend, for example, she reported, "The guy questioned us and he had his sunglasses on and he was so strict, and then he was like, 'So, what are you ladies doing after?' and I was like, 'What?' He was like, 'Yeah, where are you going? Maybe we could meet up?' I'm like, 'I'm just going to go now,' and he was like, 'Okay, have a nice day.'"

The interviewee accounts suggest border resident understandings that crossings were shaped by filtered bordering and, more particularly, that along with Indigeneity, officially perceived age, class, sexuality and gender could be significant in isolation or in combination in ways that could make crossings more or less difficult. Along with accounts of these kinds of differentiated treatment, many also commented on the significance of other officially perceived ethnoracial identities for crossing experiences. Mary recalled, for example, being instructed as a customs officer that "all the Italians smuggle, so you've got to go after the Italians," a comment that resonates with Dylan's observation that Italian-Canadian friends of his "would always have to pull over and ... show something [documents]," while his own non-Italian family "never had that problem ... it was never a question of our citizenship or [that] we were in question as people." Bill and Susan noted that their European immigrant parents had "accents," which led to more questioning at the border, and as a result, they always carried proof of their naturalized citizenship status. "My dad," said Bill, "always had to pull out ... his citizenship card," and Susan's father, a Canadian citizen for over two decades, regularly had to produce his Canadian passport.

While the interview schedule did not include an explicit reference to racialized identities, the broader question that asked interviewees "As a child were you aware that some people might have problems crossing the border?" was often accompanied by interviewer prompts to share observations regarding the experiences of "visible minorities," "people of colour," or the possible role of skin colour or "race" in border crossings. Canadian Niagara was the site of a 1999 case of racial profiling taken to the Canadian Human Rights Commission by Selwyn Pieters after he had his luggage searched by a student Canada Customs inspector on a train at Fort Erie. The settlement of the case in 2002 was supposed to involve "a ban on racial profiling at the border" as well as "a requirement that passengers be told why they are being selected for secondary inspection" and an agreement by Customs "to ... conduct anti-racism training ... [and] to collect race-based data on who is selected for inspection" (Tanovich 2006, 174–5). While this case was not mentioned by any of the interviewees, many of the predominantly white respondents described official targeting of blackness in the pre-9/11 period consistent with this case and the literature on racial profiling in Canada more generally.[4]

Before discussing this further, I note first that some of the white interviewees claimed that they had very little to say on the topic of racialized bordering, due to their lack of exposure to racial diversity. Such white claims of racial innocence were, in turn, often linked to constructions of a white borderland space implicitly or explicitly juxtaposed with the allegedly more multiracial Toronto and environs and/or the U.S. side of the river. Avery, for example, claimed that she knew little about racialized bordering because "nobody really in Niagara is ... of a different origin or whatever. Like it's not your friends that are." Barb also volunteered that as a child she was unaware of the issue, because she had gone to school with "largely white, middle-class families ... I didn't really have the opportunity to be exposed to other perspectives and opinions and experiences." Such statements suppressed the history of colonial and racialized violence that produced the putatively white geography of Canadian Niagara, while simultaneously revealing patterns of racialized segregation and what has been described by scholars of blackness in Canada as the "absented presence" (see Walcott 2003; McKittrick 2007) of the non-white populations of the region (e.g., Kobayashi 1998).

In contrast to such disclaimers, however, many other white interviewees directly introduced the topic of race at the border. Ed did so to explicitly deny any suggestion of racialized bordering, claiming that he was not "treated any differently when I was [crossing] with my black friends than my white friends." This explicit denial of a racially filtered border, however, contrasted with other interviewee accounts.

Hannah, who first suggested that male crossers might experience more scrutiny than female crossers, for example, followed this by adding the observation "[but] it's a lot worse for different races." Mary described how, while working at the border, she observed a racist "pattern of harassment around blacks" that was justified by other border officials on the grounds of allegations that they were more likely to have guns (in her experience, however, most of the guns confiscated at the border came largely from American police). As with her description of how Canadian female border officers such as herself were kept away from the "Indian" bus, she reported that they were also withdrawn from their posts when "American customs would phone over and say, 'There are a couple of black guys coming over who are drunk or who appear to be on drugs.'"

Lisa's janitorial work at one of the bridges, included observation of racialized filtering at the Canadian border as she noted, "This is very stereotypical and racial, but if there was a car of black Americans that was overfilled that would be turned away [by Canadian border officials] … we [other border workers] used to be able to … pick out who was getting pulled over." Tom recalled that one time when he was pulled over when crossing for a "big rap concert," officials were "expecting drugs to be crossing the border" and so were inspecting everyone, but then added, "If you were black you were going to get questioned more."

Zak also related that one time when he and some friends were pulled over, he was pretty sure that the U.S. border officials were looking for drugs and that their car was targeted because one of his friends was "mulatto." He concluded this, he said, because "there wasn't really anything out of whack with anyone else … [so] race hit us right away." In this story, the apparently unusual experience of heightened surveillance was attributed to official criminalization of blackness, with unnamed whiteness being the unspoken norm that made his "mulatto" friend "out of whack."

The significance of blackness was again emphasized by Sheila, who told a story about the experience of a friend's African-American boyfriend, who was "pulled in" by Canadian officials – who then called her friend's house and asked her father, "Do you know that this guy is coming to your house?" This treatment, she felt, illustrated how border crossing into Canada "really has a lot to do with your ethnic background. Like, if you're not white then they lay it into you, which I don't think is very fair." She went on to add, "I mean, he's in university, he's a good kid, he works, he loves his parents ... but because he's black he got pulled right in." Ashley also described her apprehension while crossing over the local border with a male friend who was a Canadian citizen but "ethnically ... Nigerian" with a "strong accent." Although not explicitly mentioned, these latter accounts hint at how the surveillance of young black men at the border may have intersected with the gendered regulation of young white females through heightened surveillance of "interracial" couples.[5]

There were references to racialized bordering even when explicit race talk was avoided or involved hesitant responses, as in the following exchange between Debbie (respondent) and the black Ghanaian graduate student (interviewer) about people who experienced greater hassle at the border:

Respondent: I know some people, if they look suspicious, so obviously ... [border officials] would be suspicious of them, which is not always right, but it is for safety reasons so ...

Interviewer: So you talk of "being suspicious." In what way would someone be seen as "being suspicious"?

Respondent: Well, I know, like a few people, I know it is racist and I don't believe in this, but you know, a lot of people, their colour, because they have a different colour, right? I find that, I really don't know ... [border officials] think they [people of colour] are violent ... because I remember, I had a friend growing up and she is coloured, right? And sometimes ... they'd stop her and they'd be asking her all these questions. And they won't ask me all the same questions, which I find wrong but ... for some reason, they link those [being "coloured" and violence]. I don't know.

In this exchange, the respondent suggests that those who "look suspicious" might be stopped "for safety reasons," and when asked to elaborate on this, she links "looking suspicious" with being "a different colour " – a linkage that is problematized as "racist." She goes on to describe the prolonged questioning of a "coloured" friend as "wrong," but the fractured narrative communicates discomfort with being prodded on the topic. The invocation of the "coloured friend" and repeated claims of ignorance asserted an (unnamed) "white self-innocent of racism" (Frankenberg 1993, 188).

Along with their own observations as indicated, white interviewees also referenced knowledge of racialized bordering derived through their own direct or indirect personal links to border workers. Tim described how border officials would reproduce "negative stereotypes," but distanced himself from their content when he reported, "One guy in my [university] class, he's a customs officer, and pretty much all of the negative stereotypes ... about different races crossing [the border], he was pretty much reinforcing ... I don't really want to say the stuff he was saying just because it is kind of offensive." Anna also related that she had learned from a friend working at the border about "the stereotypes that they [the Canadian border officials] ... follow, that they're like ... you see two young guys in a car, you see one black guy and some are white, or you see two black guys in a nice car and they will pull them over ... definitely."

She went on to describe waiting in line while waiting to re-enter Canada with a friend and that "she picked off every car [that was experiencing a delay at the Canadian border] and she explained to me why. And it was all stereotypes. Like she just looked at them and she knew [that they would be delayed]."

Similar to Tim and some other respondents, Anna used the language of stereotypes to signal distance from, and critique of, the filtering that she was describing. Critique, however, could be combined with acceptance, as when Lisa (respondent) described the delays of African-Americans as "stereotypical and racial," but when asked to elaborate by the interviewer, linked such treatment to their allegedly less-than-core citizenship, thereby revealing an imaginary of normative white Canadianness and Americanness:

> Interviewer: So you mentioned black Americans and the way people would look, so I am assuming that skin colour was kind of a major marker [for hassle at the border].
>
> Respondent: Yeah, just because you could tell they're not native American [or] Canadian.

Here the white interviewer is made complicit in the presumption that exclusion from "native" Americanness or Canadianness justifies greater surveillance, as racialized bordering is linked to broader constructions of racialized national belonging and exclusion. In a parallel way, Tom, who had described the targeting of black crossers on the night of a concert, had led into this example stating that he knew that "sometimes, unfortunately … certain nationalities would get a certain eye, even before 9/11 … they would be, I don't know, for lack of a better term, 'risk factors.'" The slippage here between race and nationality again constructs blackness as outside Canadianness.

While accounts of racialized filtering in the pre-9/11 period tended to focus on the disproportionate surveilling of blackness, June – a frequent crosser in this period – also recalled that "crossing either into the States or into Canada, sometimes you would see a car full of people that … appeared to be, say, East Indian, or something of that descent and … a lot of time it seems like they were delayed or pulled over."

Joe, who was of Middle Eastern origin, also reflected on how in the pre-9/11 period, even after his family became Canadian citizens, they still experienced difficult crossings. One time, for example, his family was crossing the local border to go to a wedding in the U.S.: "[We] got turned back … because my dad didn't have some kind of paper that he needed … we got sent over to the immigration office and …it turned out to be quite a hassle." Crossing independently of his family in the pre-9/11 period, Joe also found that when border officials learned of his country of birth they seemed distrustful and would ask "a lot of questions, like where I was going to school, or what I was doing, and why I was going there, whose car it was." His non–Middle Eastern Canadian friends, he noted, "never seemed to have the same kind of trouble."

These examples make clear how interviewee discussion of ethnoracialized filtering at the border often overlapped with references to nationality. Chuck, who was living in the U.S. during high school, recalled, for example, that the "biggest trouble" he encountered as a teen crossing back and forth was when he was travelling with a friend who was a citizen of India, because this friend was "always hassled." Chuck felt that "it was blatantly obvious" that this was due to his "ethnicity," because he "was a great guy [who] had never done anything wrong in his life … one of the politest guys."

Interviewees made other more explicit references to filtering by citizenship status. Richard, who attended a private boarding school in which several international students were enrolled, for example, recounted that these non-Canadian "visa kids" would be treated differently in the course of school trips that involved crossing the border:

> Sometimes they [the school] would just have, like, one bus for the visa kids. It'd be all the Mexican kids and the Chinese kids, and they'd all be on that bus. We'd see that bus pulled off to customs and they're all getting

> out, trying to show their papers and everything ... We were going across to play a hockey game and one of the guys on the bus ... he told the guy he was born in Moscow and that he was from Russia and everything, and then he [U.S. border official] just grabbed the kid and just pulled him right off the bus ... We had to wait for, like, an hour for the kid to get back.

Michael, from a working-class family in Niagara Falls, Ontario, described how during a school trip to go skiing in the U.S., the teacher told the non-Canadian students to tell border officials that they were Canadians in order to avoid "hassle at the border," and the strategy worked: "They were fine; we got through." Sophie, who had grown up lower middle class in Niagara-on-the-Lake, recalled how having her landed immigrant mother in the car made the crossing more complicated and that her mother was "paranoid" and "very nervous" about border crossings due to the intensive questioning that she experienced.

Interviewee discussion of pre-9/11 border crossings, then, make clear that they understood that filtered bordering eased the cross-border mobilities of some while hindering (to varied degrees) the cross-border mobilities of others. The negative psychological impact on those crossers who had to anticipate greater scrutiny as a result of such filtered bordering was also often recognized.

Filtered Bordering Post-9/11

Accounts of post-9/11 securitization included many references to an intensification of filtered bordering – in particular, there was considerable commentary about deepened and expanded official scrutiny of ethnoracialized Others. As will become evident below, references to filtering in the post-9/11 context were also accompanied by greater editorializing about its appropriateness than was the case for the pre-9/11 period. As already mentioned, for example, Yvonne described as newly problematic that she and her girlfriends "flew right by [the border]" because they "looked harmless," arguing that instead of this gendered facilitation "there should have been more security."

There were also some ambivalent references by interviewees to Indigenous crossings that were portrayed as problematically privileged (i.e., under the Jay Treaty) in a context of deepened and expanded border securitization. Kerrie, for example, reported that a friend working at the border had told her that border officials still weren't "allowed to question Native Americans ... no matter what, they just let them go," but also how (in contradiction to the previous assertion), if border workers did pull them over, "they would come back and protest. Huge groups of them. It would just be chaos and it wasn't worth it [to inspect them] because they … [had] so many rights." Erin reported being instructed by other border workers, "As soon as they say 'Six Nations,' just let them cross the border." Despite her own familial connections to Six Nations crossers, she appeared to be rethinking this form of filtered bordering in the post-9/11 context when she added that she "thought that was kind of weird … Why shouldn't we be asking them questions?"

More significant than the limited discussion of allegedly privileged cross-border mobilities associated with Indigeneity however, was the suggestion from white interviewees as well as the two "visible minority" Canadian citizens interviewed in this study that post-9/11 changes had intensified and expanded already ethnoracialized filtering. The accounts are consistent with reports by non-local advocates for civil liberties that the 2002 settlement of the 1999 racial profiling case did not end ethnoracial profiling at the Canadian Niagara border. (For the case of a re-entry of a Canadian of Middle Eastern descent at the Queenston-Lewiston bridge in 2008, see International Civil Liberties Monitoring Group 2010, 18.)

Pratt and Thompson's (2008) analyses of interviews with front-line border officials at a major Canadian land port of entry in 2003–4 points to how personally and institutionally derived "racialized risk knowledges" shaped officers self-reported discretionary treatment of border crossers based on attributed identities of race, nationality, ethnicity, culture, and religion. Official distinctions between "acceptable" profiling based on nationality, and "unacceptable" profiling based on race, Pratt and Thompson argue, were particularly significant as "when asked about racial profiling, officers commonly responded by talking about nationality" (Pratt and Thompson 2008, 629). This slippage made it possible for border officials to acknowledge "legitimate" profiling of nationally identified non-whites (including Canadian citizens) based on "reasonable suspicion" (Pratt and Thompson 2008; Pratt 2010). This "enabling ambiguity" also supported official denials of racial profiling based on a "narrow definition of racial profiling as an explicit, formal directive from above to target members of a particular racial group" (Pratt and Thompson 2008, 635).

Both of the "visible minority" Canadian citizens interviewed in this study described restricting their post-9/11 crossings as a result of directly experienced or anticipated profiling and harassment. Joe, who as previously mentioned, had been born in the Middle East and had experienced pre-9/11 border "hassle," found that following 9/11 it seemed "like they [the U.S.] didn't want to let me into the country. They try to look for an excuse not to let me in." U.S.-imposed "double inspections" (on both entering and exiting the U.S.) had increased his difficulties to the point where he reduced his local cross-border travel because "I didn't want to have to deal with Americans both ways."

Ameena, who had described the harassment of children "of different races" in post-9/11 Canadian Niagara, spoke about ongoing border-crossing challenges. While telling of a local professional who had not been allowed to board a plane in the U.S. because he was Muslim, she added that it was the "same thing with the [local] border" where, she had heard, U.S. border workers would now get out of their booths and ask, "'Where are you going?' ... They would ask you 'What's your name in full?' And if your name was ... [a] Muslim last name, they wouldn't let you enter." A relative who used to cross daily, she reported, now crossed only "once a year," and Ameena's more immediate family had "heard so many stories" about trouble at the border that they had stopped crossing completely.

As indicated, many of the white interviewees reported understandings of post-9/11 filtered bordering consistent with the accounts from Joe and Ameena. Three with close links to border work and/or workers were explicit in labelling such behaviours as "racial profiling." Dustin, an upper-middle-class respondent from Port Colborne who had border-worker friends, said of the immediate post-9/11 period, "Things have changed, and they are increasing security. Unfortunately, some of it is ethnic, and the ways they [border officials] identify suspects is racial profiling." Dale also learned from his border-worker friends about filtered bordering based on age and gender, but after 9/11, he had been told that "if you happen to be Arab, you'd have a hard ... time going through," and as a result decided not to cross the border with a Turkish friend in order to avoid what he anticipated would be a "huge hassle."

Erin also used the term "racial profiling" when she explained that her reasons for leaving her border-worker job included unhappiness with this practice as well as poor treatment of young people and unwarranted strip-searches. She described her personal distress at the intensive questioning experienced by a Muslim friend and critiqued how she herself was treated by U.S. border officials, who searched her car and asked what she considered to be an unreasonable number of questions when she crossed the local border to visit this friend in the U.S. Michelle did not use the term "racial profiling" but reported that an acquaintance working at the border had told her that "basically anyone with colour is going to be questioned in depth ... especially if they look of Middle Eastern descent."

Interviewees without direct links to border enforcement drew upon personal observations and the reports of friends, family, acquaintances, and the media to offer other examples of ethnoracialized filtering. Even those who purported to have little to say about "race," indicated awareness of intensified official scrutiny of those who were "different." Tim, who stated that he had little experience because he hadn't "crossed with [a] ... person of another race," went on to describe how border officials would "definitely check out other cars" where "you can sort of see that they're [the passengers are] different. It takes them a little longer [and] if you're in a wait, you can usually ... see the different kind of people in the car."

Others were more explicit. Kerrie, for example commented, "Anybody East Indian, or that looks it, they pull over, regardless." Describing a crossing into the U.S. over a year after 11 September 2001, Michael recounted, "When we got close to the actual [U.S.] border ... there was a customs officer walking down the aisle between the cars looking in everyone's car. He passed our car, but I did see [in the case of] any coloured people [or] someone wearing a religious headdress, he would stop at their window and ask them questions."

The degree to which officially perceived non-whiteness or religious "difference" was linked to greater hassle at the border also emerged indirectly in stories from white interviewees, who wondered if border officials' more intensive questioning of them was linked to their own slightly darker skin. Kerrie related that when she was taking her parents across the border to the Buffalo airport, U.S. officials asked for her driver's licence. In her licence photo, she reported, she was "really tanned and they asked if I was really Canadian. So I don't know if that's why they asked, but it was weird." While not certain that her tanned skin had led to the questioning of her nationality, she presented this as a plausible explanation for the otherwise "weird" situation of having her citizenship questioned as a clearly self-identified (but still unnamed), white crosser.

Anxiety about being incorrectly ethnoracialized as "Arab" or "Muslim" on the basis of a darker skin tone in the post-9/11 period was even clearer in the account of Jim, who suggested that "if you're Italian, having olive-coloured skin and can grow a thick beard, I've seen cases where they'll think you're an Arab ... even if you're not even close." That this was a more personalized concern was made clearer when he added the observation,

> Even if you look like you might be Muslim, now my own appearance, I mean, I'm obviously not, but I've got dark hair, a beard, somewhat of a darker, you know, tanned complexion ... so if I go over that border, I could have a problem, even though ... I don't look Muslim, but that remote chance, I mean, you could get interrogated.

Here the presumed consequence of being officially ethnoracialized as Muslim by virtue of a darker complexion was a loss of both whiteness and Canadianness and an increased risk of "interrogation."

Judith also speculated that the reason that her white son-in-law was stopped at the local border was that, although "he's a Canadian ... he's really dark ... [his mother is] French-Canadian and ... he's got dark curly hair and high cheekbones ... they ... thought he was a Dominican or something, [but] he was born in Welland." These stories reveal awareness of the contingency of racialized categories as whiteness in this case could be "misread" by border officials – as well as cost of such "misreadings," that is, eviction from a self-evident "Canadianness" and resulting constraints on cross-border mobility.

Respondents offered critical assessments of what they saw as intensified ethnoracialized filtering at the border, describing this as "unfair," driven by "stereotypes," and/or "racist" (particularly when such filtering affected friends and acquaintances), but such statements were also often accompanied by additional suggestions that such processes were nonetheless acceptable in light of the security threat.

As is evident in some of the commentary already reviewed, one of the ways in which Canadian interviewees could offer critique of border securitization without overtly undermining Canadian-side enforcement, was by focusing on the practices of U.S. officials. Sheila, who had problematized the Canadian border treatment of the African-American friend-of-a-friend, was also critical of U.S. officials' alleged treatment of a Palestinian-Canadian friend-of-a-friend. She reported:

> Even though he was born in Canada [and] is a Canadian citizen ... they [U.S. border officials] were just so mean to him because of his background that they [her friends] didn't think they were going to get over the border.

> So they [the border officials] just searched the car up and down ... and then they sent them on their way, but he [the Palestinian-Canadian] was like, "Sorry, guys, you know, like, my parents were born there," and they [the U.S. border officials] take it out on him.

Dave offered an even broader indictment of U.S. border officials' behaviour towards fellow Canadians: "I remember seeing on TV once, this story about four young men that were of Middle Eastern descent, and they were Canadian, and they were just going over the border [to the U.S.] ... and they ended up being stopped and put into a cell for four hours ... That really ticked me off because they're Canadians." The pan-racial Canadianness invoked here went beyond personalized outrage at the treatment of friends to challenge U.S. racism and, more indirectly perhaps, constructions of Canadianness that excluded those of Palestinian or Middle Eastern descent. Attributing ethnoracialized surveilling exclusively to American authorities in these cases, however, deflected attention from similar behaviour by Canadian border authorities and fed easily into well-established tropes of Canadian racial progressiveness or even racelessness relative to the U.S. (Tanovich 2006, 504).

As indicated, while some critiqued such practices, interviewees also often offered (sometimes simultaneous) support for ethnoracialized filtering as a necessary part of post-9/11 securitization. Kerrie, who had worried that her tan had resulted in a questioning of her Canadian citizenship and described U.S. border practices as "discriminatory," nonetheless accepted such practices as reasonable, given post-9/11 U.S. concern that Canada was a source of terrorists: "There's definitely that discrimination, but at the same time, I think they [U.S. border officials] kind of had to, because I guess Canada was their main concern at that time, and even though it was maybe a little discriminatory, I think they had to do it that way."

Miles also labelled U.S. actions as "unfair" but nonetheless justified when he commented:

> American customs has really been cracking down on anyone that really isn't Caucasian, which I guess I can understand to a degree because of the problem with the terrorists ... but it's kind of not fair because just because you look a certain way doesn't mean you have ... [a terrorist] belief system ... You have to check those people, but it's not fair because not everyone's a bad person, just because you're of a certain nationality.

Here the racialized category of non-Caucasian slipped into those of a "certain nationality." The race-evasive language of nationality then positioned non-Caucasian racialized minorities as less-than-full or non-Canadian or U.S. nationals who could be legitimately profiled. Miles also drew a parallel between racialized anti-terrorist "cracking down" and more routine cases where the racial profile of an escaped criminal would lead border officials to be reasonably "scrutinizing everybody of that nationality" – again collapsing race and nationality. Of this latter kind of targeting (and, by extension, the more general anti-terrorist measures), he suggested, "It's not fair, but it's just a security thing."

Dustin, who stated that increased "prejudice" was "becoming a fact of life that people have to deal with," considered "racial profiling" to be justified on the grounds that "if you weren't a criminal and you didn't have any reasons … to … hide then it shouldn't be a problem." Likewise Debbie, who had described as "wrong" the way that her "coloured friend" was subjected to greater questioning than she was before 9/11, also claimed of the post-9/11 period, that she hadn't "heard … many stories of innocent people … having a problem at the border."

While an interracial friendship had provided Debbie with insight into the negative impact of ethnoracialized filtering, the cost for her friend was nevertheless downplayed in the suggestion reproduced above, that few "innocent" people experienced border "problems." Ethnoracialized crossers, clearly acknowledged by many white interviewees to have more difficult border crossings, could through this circular logic, be understood as less innocent as a result of being more intensely surveilled. These statements then reveal how recognition, and in some cases, even partial critique of filtered bordering by white interviewees, could coexist with legitimation of disproportionate surveillance and accompanying violation of the rights of ethnoracialized crossers.

Chuck, who had critiqued the pre-9/11 treatment of a friend who was problematically hassled due to "ethnicity," when reflecting on why people would experience trouble at the post-9/11 border, suggested that this could stem from "their attitudes at the border towards the guards," "their past history due to crime," or "the colour of their skin," because since 9/11 "people of Middle East, Asia [sic] ethnicity do have a lot more difficulty crossing the border than they did before." The list was followed by his more editorial commentary, "Whether it's right or wrong, I can understand where the border guards are coming from."

Judith, who shared the story about her "very dark" son-in-law being stopped at the border, was also largely supportive of greater border security. As she explained, it was to be expected that the border would be "more strict" post 9/11, but she added: "You're presuming it's not going to affect you. As [a] Canadian or as someone local ... you feel you have a right to go over wherever you feel like it ... but anybody looking different I presume they'd want to check [at the border]."

It was when she was pressed to clarify whether this reference to people "looking different" referred to racialized minorities that she responded with the story about her "dark" son-in-law; but when asked by the interviewer whether her son-in-law's experience made her think about "the experience of people who are Canadians that are darker," she answered with a lukewarm "Yeah, I suppose," but then suggested that border officials might still be too "lax" given the security threat. The attempt on the part of the interviewer to link the discussion of her son-in-law to ethnoracialized filtering more broadly, and to offer her an opportunity to reflect more critically on these processes, did not appear to unsettle her unstated assumption that Canadians and locals were white and that it was reasonable that those who "looked different" would be more intensely surveilled.

Attempts to reconcile critique and legitimation of ethnoracialized filtering were apparent in efforts to distinguish between acceptable criminal profiling and racist behaviour on the part of border officials. Kerrie, who thought that her tan may have prompted questioning about her citizenship, used such a logic in a description of how African-Americans driving a "nice car" got "pulled over a lot more than a white person would" when entering Canada. She suggested that this practice (based on an unstated but presumptive disjuncture between the class marker of owning a "nice car" and racialized blackness) was security driven, because while "a lot of times ... [the border guard] could be wrong ... that one time that they're right [they] could save someone maybe." When asked to clarify how the possibility of "saving somebody" made such targeting legitimate, Kerrie first appeared to problematize such practices but then invoked the motivation of the Canadian guard as central to determining whether the action was racist or not:

> Maybe so not much an African-American driving a nice car. I mean it shouldn't just make up ... [a border official's] mind like that. They should still question them or whatever. But I guess it depends on the person. If [border officials] are doing it to protect the country, then it is one thing, but if they're doing it because they're racist, then I think it's totally wrong.

Kerrie here asserted that racism was wrong, but separated racism from Canadian border officials' practice of stopping African-Americans in "nice cars," which, she suggested, might be done to "protect the country." What might otherwise be deemed racist and therefore "wrong," was then justified in this account, by a security rationale and was portrayed as therefore not racist. The determining factor as to whether such selective stopping was racist or not was allegedly the (externally inaccessible) interior motivation of the border worker (i.e., whether actions were motivated by the desire to protect the country or by anti-black racism).

One of the more striking examples of ambivalent support for ethnoracialized bordering came from Ameena, who critiqued the profiling of those with Muslim names while appearing to support greater border securitization. She noted, for example, that it was still "so hard to get over the border. They will check your car continuously... when you cross and [are] coming back." But she then added, "The border's doing a really great job, because if someone does go into their country [the U.S.] and does ruin their country ... no one has the right to do that ... what they have at the border, it is good, to protect their country from other people." When pushed by the interviewer to clarify whether intensified surveillance was problematically discriminatory towards Muslims, she responded by saying, "Well, I know that it is discriminating ... because not all Muslims cause the problem." She then suggested, "We [Canadians] shouldn't be discriminatory [at the border] because ... you can't discriminate [against] anyone without a reason," but added that Americans had a reason to discriminate "because ... [of] September the 11th." Later in the interview she further reflected, "Freeing the border wouldn't be that great because anyone can enter the States or Canada, and then if something goes wrong, then we're in trouble ... that's why they [border officials] should be a little bit strict, and a little bit free." Here, securitization involving the profiling of Muslims appeared to be supported as a necessary protection, although whether this was against global terrorism or perhaps anti-Canadian or anti-Muslim discrimination was unclear.[6]

While some of these border residents critiqued ethnoracialized filtering at the post-9/11 border, then, many also positioned ethnoracialized Others as suspicious and potentially threatening to a presumptively white country and people whose protection, it was suggested, required such practices. As Razack (2009) notes, this demonstrates the importance of "following the colour line behind security discourses, [of] tracking the racial logic that divides humanity into those who are threats and those who must be protected" (819). Interviewees' discussion of ethnoracialized bordering that portrayed this as an acceptable aspect of post-9/11 securitization often did so through nationalized constructions of inclusion and exclusion that drew upon and reproduced "racialized structures of citizenship in which people of colour, suspected of duplicity," are "policed and kept at the margins of law and community" (Razack 2010, 89). The discussion of filtering at the post-9/11 border points to the differentiated impact on diverse Canadian Niagara border residents. The narratives also illuminate aspects of borderline nationalism. I begin to address the latter in the section below on dual citizenship and then take it up more fully in chapters 5 and 6.

Dual Canadian/U.S. Citizenship

As indicated earlier, there were some references in the interviews to the experiences of those who were neither Canadian nor U.S. citizens. Here I extend the discussion of citizenship at the border to interviewee descriptions of how dual Canada/U.S. citizenship was experienced and perceived at the borderline. I have separated out this discussion because of its particular significance to a discussion of filtered bordering and national identity in a border region marked by a significant degree of everyday border crossing. What is striking is the perhaps counter-intuitive finding that while dual citizenship was considered an asset in terms of broader binational economic opportunities, it was also linked to more fraught pre- and post-9/11 border crossings as well as sometimes uncomfortable positionings within a wider regional border culture. Here I focus on the former, leaving discussion of the latter for the next chapter.

Five of the interviewees (June, Dylan, Rob, Ashley, and Kevin) were dual citizens, and Jim was working towards this status. In addition to these individuals, another three (Jim, Sheila, and Tom) had parents who were either U.S. or dual Canadian/U.S. citizens. Accounts from these individuals suggested that while dual citizenship was understood to convey potential benefits by opening up access to economic opportunities on both sides of the border, it was also experienced as a liability in the border crossing itself and as a result was often concealed during border inspections at Niagara. Sheila, for example, described learning that her U.S.-born mother's dual citizenship status meant that, when her mother crossed back into the U.S., "she's only accepted as American" but could be recognized as both "Canadian and American" upon her return to Canada. In an attempt to avoid hassle, her mother stated only her Canadian citizenship for both trips, a strategy that Sheila described as having its own inherent risks, because "if they [U.S. border officials] ever found out she was dual and wasn't saying 'American' she could get into a lot of trouble."

Dylan, who had dual citizenship, described his own childhood confusion engendered by what he understood to be a requirement to present as a mononational at the local border. He described how his father would state his citizenship as "Canadian," and his mother would say "United States," but in his own case, "I didn't know how to reply ... I would just stutter ... and my mom would be like, 'Oh, he's a Canadian,' and my dad would be like, 'Oh, he's an American,' which was funny because ... [their nationalities are] the opposite."

Dylan also related a story about the challenge of declaring citizenship in his first independent encounter with U.S. border officials:

> What happened was, he [the U.S. border official] goes, "What's your citizenship?" ... I said "American." And the guy just went off ... like, "What do you mean, American? North American, South American, Central American?" And ... this is my first time doing it by myself and I don't have my parents to protect me or anything so I'm like, "North American." He's like, "Is that Canadian, United States?" And I'm like, "United States," and he's like, "Then why do you have Canadian licence plates?" And he just started going off like that and I didn't know ... what's going to happen? Like, am I going to get deported or what?

After this experience he declared only his Canadian citizenship when travelling with friends "because I didn't want them to get stuck up in this hassle." Kevin likewise reported that "you don't declare the dual citizenship, I just say 'Canadian,'" while Rob explained in greater detail how after experiencing a "bit of trouble because a lot of border guards don't really recognize dual citizenship," he learned that it was best to declare that he was American when crossing into the U.S. and Canadian when crossing into Canada.

Even declarations of mononationality, however, did not necessarily reduce hassle for Canada/U.S. dual citizens. When crossing into the U.S. for snowboarding, reported Tim, a friend with dual citizenship "would be the only one who would say 'American' and [the friend] would ... get a second look ... [be] questioned, check the ID." Most accounts of troubled entries for dual citizens related to U.S. entries revealing asymmetries in bordered processes and preoccupations. June emphasized this difference when she highlighted how both her childhood and young adult entries to Canada were easier than U.S. entries and how at the time of the interview, "a couple of the customs agents of the Canadian side ... [would] recognize [her] from crossing quite a bit and just [say], 'Oh, you going to school? Okay, bye. Have a good day.'" Overall, however, the best strategy for border crossing was agreed to be that of making dual citizenship invisible at the border site. The apparent challenges associated with dual citizenship at the border provide a striking example of the limits of official regional binationalism as the border was experienced as unable to accommodate the everyday juridical realities of those who, on the surface, most closely embodied such binationalism – notably Canada/U.S. dual citizens living in Niagara.

Privileged Mobilities

As the previous discussion reveals, interviewees were able to offer many examples of filtered bordering in the pre- and post-9/11 periods. As I have also emphasized, some of this knowledge came from direct or indirect links to Canadian border workers. That young Canadian Niagara border residents were also socialized as children into assessments of "safe citizenship" that paralleled border workers' filtering practices was apparent from some recollections of how their parents might make negative judgments about those observed to be experiencing delays. Tim recalled, for example, how when officially produced delays of other crossers slowed his own family's travel, his father would "mumble some racial slur of some car in front of us," while Jessie's father would comment on those experiencing prolonged questioning, saying, "Well, look what they're driving; they look like troublemakers." Kevin's father also said things like, "Look at that guy, driving a beat-up car, he's got long stringy hair, he kept his sunglasses on, he might have been smoking a cigarette, just a lot more suspicious." Such statements about allegedly predictable markers of "trouble" and "suspicion" reveal how young border residents learned about and came to understand filtered bordering.

Within this context, there were suggestions of how they and other border residents also worked at successfully navigating the border though honed performances aimed at communicating "safe citizen" status. Many expressed pride in their ability – as allegedly demonstrated by eased border crossings – to communicate such status. Chuck, for instance, attributed his family's ability to cross easily into the U.S. during a period of "high alert" associated with the Persian Gulf War, to the fact that "we were very polite, we always had our ID ready. We treated them [border officials] with respect and after a while they got to know us." While such border performances were credited with easing the crossing, he also believed that official designation of his family as "safe" was probably also confirmed by the surveillance technologies of state databasing as he guessed. "They had information … on the computer which knew how often we crossed … the reasons we were crossing, [and] thus after a while, they would ask us, 'Citizenship? Where are you going? Have a good day,' and we'd be on our way."

Difficulties encountered by family or friends at the border were, then, also understood as anomalous "misreadings" by border officials of those known to the interviewee to be "safe" (including those acknowledged as occupying officially perceived categories of risk). Through such processes, interviewees could still offer a degree of critique of any hassle experienced by themselves, family, or friends at the border site without necessarily challenging the overall processes of filtered bordering. Interviewees then offered only limited challenges to the filtered bordering processes that they described, with the striking exception of some problematizing of allegedly privileged Indigenous mobilities.

A flip side of this was that the privileges associated with being officially perceived as "safe" were not explicitly acknowledged, let alone troubled. An example of how this worked comes from Dustin's reference to a border-related story that "everyone" had heard "about the old couple in the Winnebago" that was "ripped apart" to reveal "a kilogram of cocaine that someone [had] taped to the bumper." In this story, dominant constructions of those deemed "safe" crossers were reinforced as implicit in the apocryphal tale was the "safe citizen" status of presumptively white, middle-class, heterosexual, senior leisure travellers. That "everyone" had heard about such crossers being inadvertently implicated in criminality (precisely because of being equally self-evidently less likely to attract sustained surveillance), reproduced constructions of safe and unsafe crossers while simultaneously upholding the importance and authority of border inspection (including of "safe citizens").

A partial exception to the silence regarding the structural privilege attached to being read as "safe" and its links to race and class in this context was evident when Michelle, in the course of reflecting on her easeful cross-border mobilities prior to 9/11, speculated vaguely on the possible role of racial and class privilege, saying, "I don't know, maybe it's just because I'm white and we're middle class." More indirect acknowledgment of racialized privilege emerged in the stories about official "misreadings" of whiteness as non-whiteness, and in expressed concerns that post-9/11 securitization might be marked by more universalized (e.g., less discretionary) processes. As discussed, some white interviewees were angered by the possibility of being subjected to some of the same constraints (e.g., more prolonged questioning and greater possibility of being searched) that they recognized had long been experienced by ethnoracialized Others. That universalized border practices might result in more effective border enforcement is suggested by Pratt's (2010) reference to the claim of one senior manager with the Canada Border Services Agency that "statistically generated randomized searches" generate "more 'hits' (searches that result in a seizure)" than "case-by-case risk assessments by border officers" (470). Such an outcome seemed to be exactly what Lisa was anticipating when she expressed anxiety about locals with undeclared goods being "caught" by post-9/11 securitization.

White interviewees, as mentioned, shared their frustration with an anticipated loss of eased crossings due to less discretionary, more universalized forms of border securitization. Dan, whom I quoted earlier complaining that locals like him were being negatively impacted by post-9/11 U.S. border securitization, described how the new possibility of being given a "hard time" made him "kind of angry" and affected his previous practice of regular border crossings, because, as he put it, "I don't like going there [the U.S.] anymore. I just want to get out as fast as I can."

The suggestion by white respondents such as Dan that the surveillance associated with post-9/11 securitization was new and was affecting all local border residents equally, however, obscured both the history of filtered bordering (and especially ethnoracialized bordering) prior to 9/11 and the disproportional surveillance of ethnoracialized border residents in the post-9/11 context. Sheila, whom I quoted earlier critiquing Canadian border enforcement's treatment of her friend's African-American boyfriend, for example, complained about the ostensibly newly intrusive bordering of the post-9/11 period when she stated:

> Everything started to get really harsh. The Americans are brutal at the border, and they just look you up and down, and they want to know

> everything about you and where you were? And how much you spent? And why you were there? And why you couldn't buy that in your own country? And why do you have to come here?

While aware that border crossing had been "harsh" for ethnoracialized others well before 9/11, she was nonetheless frustrated at the perceived loss of her own (unnamed) racial privilege post-9/11 and expressed her desire for a new form of visibility, "a sign," that would ensure continued facilitated mobility: "I just want to wear a sign that says, like, 'I have no drugs, I have no weapons, I go to university ... I'm not a bad person, you can trust me, I've had a job for five years ... the same one' ... I should have a list across my shirt that just says 'Okay, I'm clean ... let me go through!'"

Despite concern among some white interviewees about the possibility of more universally applied border securitization, their accounts also suggested that there was, in fact, little if any diminishment of the relatively privileged positioning of whites within post-9/11 filtered bordering processes. On the contrary, it was indirectly acknowledged that those officially perceived as white were continuing to benefit from relatively eased cross-border movements in the post-9/11 era.

Eased cross-border mobility, in turn, translated into a structurally derived "whiteness-as-materiality" (Andrucki 2010, 360), insofar as those officially constructed as white (and other privileged positionings) benefited materially from their greater access to cross-border shopping and the related activity of minor smuggling. What Du Bois termed the "wages of whiteness" derived from "privileged status in the eyes of authority" (Winant 2004, 62) was also augmented by facilitated access to cross-border recreation, education, and employment, along with the ability to participate in and maintain cross-border families, friendships, and communities.

Relatively eased post-9/11 cross-border mobilities also reproduced an ontological sense of inherent desirability as officially perceived "safe citizens," as well as security in relation to ethnoracialized crossers who were disproportionately subjected to sustained or intrusive interrogations, physical searches, arrests, as well as (as has been pointed out for Canadian borders more broadly), life-threatening forms of detention, deportation, or rendition (Razack 2010; International Civil Liberties Monitoring Group 2010). The material, social, ontological, and security benefits of whiteness as a "passport of privilege" (Kalra et al. as cited by Andrucki 2010, 360), however, was not explicitly acknowledged, as interviewees instead portrayed filtered bordering as a mostly reasonable outcome of official assessments of crossers. The result was that existing inequalities of the border region and beyond were reinforced rather than challenged, as largely unproblematized filtered bordering continued to ease the cross-border mobilities of some while troubling those of others.

Conclusion

Chapter 2 revealed how the relatively eased cross-border mobilities of predominantly white, working, and middle-class Canadian border residents facilitated access to a wide range of economic and recreational opportunities as well as cross-border families, friendships, and communities. That and some of chapter 3 also began to point to how age, class, and gender could shape childhood and youthful crossings in differentiated and unequal ways. In this chapter, I took up the topic of filtered bordering and its effects more fully and expanded the documentation and analysis to include the references to not only age, class, and gender but also Indigenous, ethnoracialized, and citizenship positionings. The largely white interviewees had a lot to say about the everyday realities of both pre- and post-9/11 filtered bordering, suggesting both continuity and change in terms of filtering practices and patterns. The fraught juridical positioning of Canada/U.S. dual citizens at the border was an important indicator of the limits of everyday binationalism in this context. While descriptions of filtered bordering could include critique of border officials, the accompanying interviewee commentary often reconciled such critique with tacit or more explicit suggestions that filtered bordering was a necessary part of ensuring personal and national security.

Narratives of pre- and post-9/11 border crossings, then, suggest how these were shaped by a changing political economy and filtered bordering that produced unequal local cross-border im/mobilities, which, in turn, contributed to inequality in Canadian Niagara and beyond. The inequalities linked to border crossings, as reported by these interviewees, augments existing scholarship on Canada/U.S. bordering as well as the broader discussion of mobility and inequality. In the context of

Canadian Niagara, filtered bordering reproduced material and other inequalities among border residents, as some continued to enjoy a range of cross-border activities, irrespective of fluctuating exchange rates, while others were much more constrained. Such inequalities challenge dominant and popular constructions of a benign Canada/U.S. border as well as suggestions of homogenized border identity or culture.

Chapter Five

Everyday Nationalism at the Borderline

Introduction

Anzaldua's (1999) and Rosaldo's (1993) writings on the Mexico/U.S. border region highlighted how borderland areas can "constitute a transition zone," where "people residing in the same territorial or cultural space may feel a sense of belonging to either one of the two sides, to each of the two sides, or even to some hybrid space" (Newman 2011, 37). This and Vila's (2000, 2003, 2005) ethnographic documentation of how border identities may articulate with classed, ethnoracial, and gendered lines of difference and inequality – prompted my interest in border identities and everyday nationalism in Canadian Niagara.

As indicated previously, many scholars of Canadian nationalism emphasize the importance of the American Other to Canadian national identity (e.g., Mackey 1999; Kymlicka 2003; Winter 2007), making the everyday nationalism of those living right at the Canada/U.S. borderline of particular interest. Recognizing the border as a site of both binational and global flows, I wanted to examine how border resident Canadianness engaged with both Americanness and a wider globalization (Kymlicka 2003). In both cases, I wanted to consider how borderline Canadianness could be linked to the wider scholarship on Canadian nationalism, which emphasizes how it often reflects and reproduces white settler colonialism, as well as ethnoracialized, gendered, and classed inequality. In this chapter I begin by focusing primarily on how everyday nationalism in Canadian Niagara engaged Americanness, leaving the discussion of globalization for chapter 6.

I approach the topic of everyday Canadianness in relationship to Americanness by first returning to the top-down official discourses of cross-border regionalism mentioned in earlier chapters. After documenting how more elite cross-border regionalism weathered the apparent re-bordering of 9/11 securitization, I turn to more everyday understandings and imaginings of border space and identity. My discussion begins with an in-depth look at varied local expressions of nationalized cross-border similarities and differences before considering how these were often classed and racialized. After considering the implications of the latter for understandings of border identity and culture, I move on to a detailed examination of the expressions of anti-Americanism often embedded in narratives about tourism work. I conclude by looking more specifically at the accounts of dual Canada/U.S. citizens, whose experiences and perceptions appear to illustrate the limits of binational hybridity in Canadian Niagara.

Official Cross-Border Regionalism

As discussed in chapter 1, Niagara's regional history has been marked not only by nationalized military and other tensions but also by efforts on the part of some to promote what I call official cross-border regionalism, including, for example, the "international city" vision articulated at the 1927 Peace Bridge opening and the post–Second World War proposal to house the United Nations headquarters on an island in the Niagara River.

In the context of free trade, Sparke (2006) notes for the western Canada/U.S. border region of Cascadia, business communities on both sides of the border developed forms of official cross-border regionalism linked to broader frameworks of continental integration as they began "to reimagine their local regions as newly 'borderless' business gateways and development hubs" (159). In Niagara, some past and more contemporary political and business leaders have also promoted cross-border cooperation as the way to improve the region's position as both a global tourist destination and a major corridor for continental trade (Meyers and Papademetriou 2001). Despite pressures towards re-bordering after 9/11, many continued to promote such goals by harnessing them to the new securitization.

As Heyman (2010) points out for the Mexico/U.S. border, however, "border elites ... seek to maintain and facilitate [binational] economic and social relations" even while "their home country's government and politics is ... often the source of their elite status and power" (23–4). There are similar tensions in the varied pre- and post-9/11 cross-border regionalist discourses of local elites that I found reproduced in *Niagara Falls Review* reporting on border-related issues. In 1999, for example, cross-border cooperation was identified as key to achieving the goal (voiced by the mayor of Buffalo) of having "the cheapest, fastest border in North America" ("Transportation Woes A 'National Crisis,'" *Niagara Falls Review*, 2 November 1999, A9), and a year later, reporting focused on how the region's competitive position within North American economic space relied upon improved highways and bridges that could handle increased cross-border trade.[1]

In the same period, however, some of the challenges and contradictions of official cross-border regionalism in the context of asymmetrical bilateralism were apparent in the negotiations over the expansion of the Peace Bridge, a fraught process that came to a complete halt in 2000 before being restarted from the beginning. In the face of such difficulties, many local leaders worked to maintain and promote cross-border regionalism. Thus, for example, the general manager of the Niagara Falls Bridge Commission, in announcing construction on a new truck lane at the Lewiston-Queenston Bridge, reiterated the need to "keep the truck traffic in Niagara" (versus other Canada/U.S. border crossings) and added, "We have to start thinking and acting as an international zone: We can't afford parochialism" ("L-Q Bridge Gets Third Truck Lane," *Niagara Falls Review*, 31 May 2000, A1). The same injunction was repeated in the spring of 2001, when the Canadian Consul General in Buffalo told the Niagara Falls Chamber of Commerce, "To attract tourism and more economic development, you need a thriving region. You need to get beyond the parochialism" ("Business Groups Told Border Should Be Invisible," *Niagara Falls Review*, 1 March 2001, A6).

One example of an attempt to combat the region's alleged parochialism and to cultivate what was portrayed as an allegedly inherent but not yet sufficiently realized binationality was the 2000 formation of the Binational Niagara Tourism Alliance (later renamed as the Binational Economic and Tourism Alliance). The alliance was initially linked to a plan to promote both sides of the Niagara River as a tourist destination under the rubric of the "Two-Nation Vacation" ("It's Time for a Two-Nation Vacation," *Niagara Falls Review*, 10 February 2000, A6).

The effort to develop tourism based on a "one region, two nations" model was endorsed in a *Niagara Falls Review* editorial that, in an indirect allusion to asymmetrical Canada-U.S. relations, reassured its Canadian readership that cross-border regionalism did not mean that "we'll become Americanized, just that we'll work with our neighbours towards common goals" ("They're Playing Our Tune," *Niagara Falls Review*, 2 March 2001, A4). The mayor of Fort Erie, meanwhile, described cross-border cooperation in tourism as the natural outcome of "a common history, a common language, a common culture and common values," and the president of Niagara Falls Tourism commented that borders "are put on a map so we know who picks up the garbage and who does the policing" but are less important to tourists "looking for a destination experience" ("Cross-Border Effort Puts All Niagara on Tourism Map," *Niagara Falls Review*, 23 June 2001, A3).

In an article about the growing tourist industry of Niagara Falls, Ontario, and the deindustrialization of Niagara Falls, New York, that appeared in the week before 9/11, the mayor of Niagara Falls, New York, was quoted discussing her vision of placing customs outside of the two cities in order to create a borderless Niagara. She was reported to have stated that "we're really one country – we're so much alike. It would be the country of Canada and the U.S. – a tourist area without the customs booths that clutter the natural beauty." In support of this vision, she offered a more personalized binationalism when she claimed to "feel just as Canadian as I do American," adding that she had relatives in Canada who felt the same way. That this "one country" version of cross-border regionalism was, however, pushing the limits of binational possibility was suggested by *Niagara Falls Review* reporting claiming that neither country was likely to support the envisioned "Niagara Americanada" ("A Tale of Two Cities," *Niagara Falls Review*, 4 September 2001, A1).

Following 9/11, as U.S.-initiated re-bordering was combined with Canadian-side official and everyday expressions of cross-border solidarity, the contradictions between nationalism and binationalism were again apparent in *Niagara Falls Review* reporting. At the end of October 2001, for example, a *Niagara Falls Review* editorial appeared to offer a nationalized response when it urged Canadian government resistance to U.S. pressure for a "border-protection plan that would encircle the continent" (a security perimeter that was also being touted by some Canadian business leaders), emphasizing instead that Canada was an "independent country," which, in agreeing to such a plan, would "risk losing [its] own right to control [its] borders" ("Don't Rush to Lock the Border," *Niagara Falls Review*, 31 October 2001, A4). A re-bordering impetus was also voiced by U.S. officials, notably the chief inspector at the Port of Buffalo, who responded to concerns about bridge gridlock voiced by the mayor of Niagara Falls, Ontario, by saying, "People have to realize they are coming into a different country" ("Ease Border Restrictions: Thomson: Fears 10-Mile Traffic Backlog," *Niagara Falls Review*, 16 March 2002, A1).

Faced with reduced and delayed border crossings, however, other local political and business leaders worked to maintain the vision of official cross-border regionalism – distinct from, but still congruent with, the shifting priorities of the U.S. and Canadian national centres. An effort to accommodate the heightened emphasis on securitization within a "borderless" discourse, for example, was evident when the district director for Canada Customs and Revenue Agency in Fort Erie, in the context of a ceremony with U.S. Customs Services to mark International Customs Day, stated, "When it comes to the protection of our communities, or keeping our streets and families safe, there is no border and there are no boundaries" ("Border Cooperation Lauded: Customs Contributions Marked at Peace Bridge Ceremony," *Niagara Falls Review*, 30 January 2002, A3).

Similar assertions were offered over a year later, when the general manager of the Niagara Falls Bridge Commission, speaking at a luncheon organized by the Niagara Falls, Ontario, Chamber of Commerce, reassured listeners that Niagara border cities and bridges would "weather the storm" created by divergent Canada-U.S. positions on the Iraq war, because their local cross-border relationship was based on a "personal 'people to people' relationship" rather than "heads of state or government" ("Border Myths Addressed By Bridge Official," *Niagara Falls Review*, 10 April 2003, A3).

Early in 2004, a *Niagara Falls Review* editorial appeared to counsel acceptance of the more securitized border when it described holiday traffic delays (from a U.S. orange alert) as "the price we're all paying these days to be kept as safe as possible from a potential terrorist threat" ("Holiday Security Threat Status a Reminder of Price of Freedom," *Niagara Falls Review*, 2 January 2004, A4). Later that same year, a report on a binational meeting of U.S. and Canadian border mayors noted that they were addressing "the challenges of living along the Canada-U.S. border" by focusing on "infrastructure improvements, border security and bi-regional marketing" ("Border Mayors Meet With Salci," *Niagara Falls Review*, 14 May 2004, A3).

As I discuss in the conclusion, in the years following the end of my interviewing, some local U.S. and Canadian leaders continued to emphasize the importance of cross-border regionalism in the face of apparent border "thickening" (Macpherson and McConnell 2007). In

2007, a report from the University at Buffalo–based Regional Institute advised both sides of the river to "harness the advantage of being a binational region in a globalized world," where "regions with strong identities and perspectives that transcend international borders are likely to hold an invaluable competitive edge" (Regional Institute 2007, 4). An analysis of interviews with leaders of seven Niagara cross-border organizations in 2009, however, suggested that despite "extensive and intimate" cross-border ties, cross-border organizational initiatives remained "fragile" (Eagles 2010, 380).

Emblematic of this fragility was the end of a binationally organized cross-border Friendship Festival that had operated from 1987 to 2004 but in 2005 returned to separately organized Canada Day and U.S. 4th of July celebrations after the loss of funding from the City of Buffalo and New York State (Eagles 2010, 387–8). While the festival continued under the same name on the Canadian side, when I participated as a festival volunteer in 2006, I noticed that explicit binationalism appeared limited to some decorative Canadian and American flags and red-white-and-blue bunting. During the headline concert on Canada Day, the bandleader's injunction to say "hello to Fort Erie" was greeted with loud cheers, but his subsequent directive to send greetings to "our [U.S.] friends over the river" received a noticeably muted response from the largely Canadian crowd.

Some local leaders, meanwhile, continued to problematize an allegedly weak cross-border regionalism, portraying this as the product of both central, state-imposed factors including post-9/11 securitization, and "more vexing" local issues such as "cross-border differences in cultures and attitudes" that needed to be better accommodated by local residents and organizations who, it was suggested, should more consistently "embrace the cross-border nature of the Niagara region" (Eagles 2010, 389). Putatively insufficient cross-border cooperation was, then, blamed in part on "border cultures" that were constructed as not only "different" on each side of the river but also problematically *in*different or even resistant to official cross-border regionalism.

Within local elite discourses, then, there were often contradictory constructions of the informal cross-border economic and social ties that I have described as everyday binationalism in earlier chapters. Sometimes, for example, these were invoked by local leaders as providing an important "organic" basis for the more formal forms of cross-border economic and/or political collaboration being advocated. At other times, however, local border residents were chastised for supposedly insufficiently binational orientations, which were allegedly thwarting maximization of the economic potential of the cross-border region. Below I turn from these more elite forms of regional discourse to consider the more everyday experiences and imaginings of borderline Canadianness that emerged from my research.

Everyday Border Space and Identity

Material for the analysis offered below was found in different parts of the interviews, including in responses to questions about the immediate other side of the border, the significance of growing up in a border region, and border identity (see the appendix). As will become clear, many of the responses discussed how Canadian Niagara border residents could be compared to both other Canadians living farther inland from the border as well as to U.S. nationals on the other side of the river. Both were predicated on largely unproblematized constructions of a territorial border that created the "two sides" of the Niagara River as distinct nationalized spaces.

This framing erased the history of Indigenous dispossession and displacement that produced territorially demarcated Canadianness and Americanness as self-evident forms of nationalized identity despite, as I have noted, a history of Indigenous protest of settler nation-state bordering made visible in local press reports and mentioned by some of the interviewees. Because Indigenous contestation of the border was not contextualized by the local press or interviewees within past and present colonialism and Indigenous sovereignty, Indigenous protestors were, however, constructed as exotic and/or illegitimate and dangerous outsiders in ways that reinforced rather than challenged the Canada/U.S. boundary and/or nationalized settler ideologies, practices, and identities (for other regions, see Wilkes, Corrigall-Brown, and Ricard 2010).

Within this context, as I describe below, the accounts of interviewees included varied and sometimes contradictory constructions of the nationalized "two sides" of the Niagara River. A minority, for example, suggested that both sides were "pretty much the same" or that Canadian residents were "Americanized"; by contrast, most portrayed the territorial borderline as congruent with nationalized difference. In what follows, I provide a sense of some of the varied constructions of bi/nationalized space and identity offered by Canadian border residents, keeping in mind that some described their understandings as having changed through time. Melissa, for example, commented that while the two sides had seemed "blurred" during her childhood, this was no longer the case, and others also suggested more specifically that the more intensive border inspections of the post-9/11 period had made the U.S. seem more foreign than before.

"Pretty Much the Same"

I begin the discussion with the minority view among the interviewees that both sides of the river were largely indistinguishable. A number of those espousing this position had significant family ties to the "other side." Dual citizen Kevin, who grew up with an American parent who commuted to work in the U.S., for example, recalled how childhood border crossings were "not a big deal at all ... I didn't even really think of it as crossing into another country. I just thought we were going over to Buffalo." Tom, who also had an American-citizen parent, suggested that the other side "seemed like it was another Canada ... the only difference was that we had to cross this bridge to get there ... the people all seemed the same, we all spoke the same." Lisa, whose uncle and aunt were in what she referred to as a "mixed" Canada/U.S. marriage, suggested that "they should just make Fort Erie [her home town] a Canadian-American town, 'cause that's what it's like anyway."

Such comments were not restricted to those with cross-border kinship links. Emma, who crossed primarily for shopping and recreation as a child, also commented that at that time "it didn't ... register ... sometimes that you were in a completely other country." "The differences," said Tim, "weren't really significant ... a lot of things are just the same as over here ... there shouldn't really be a border in that sense. We're all just people; we speak the same language [and] except for the currency we're pretty much the same."

Erin, who had spent a lot of time on the U.S. side shopping and attending events at an Aboriginal cultural centre in Buffalo, described how "when you go over it doesn't seem as if you are in another country." When asked to elaborate, she acknowledged that crossing the border meant "being interrogated and stuff," but because the other side was "so close" and because "you have people over there that you talk to, [and have] developed relationships with, it doesn't feel like 'Oh my God, I'm in another country.'"

While such constructions echoed some of the discourses of official cross-border regionalism, the only interviewee who seemed to more explicitly espouse the latter was Jason, who had crossed daily as a child to attend school in the U.S. He felt that because both sides of the river shared "many important issues," such as international bridges, trade, and cross-border shopping, Niagarans on both sides of the border were unique in being "more international" than other Canadians or Americans.

"Americanized" Canadians

A variant of the "pretty much the same" construction was apparent when some interviewees claimed that Canadian border residents were "Americanized." Jill, who had lived in the U.S. for six months as an adult and was interested in returning there, for example, reported that she didn't think of the U.S. side "as another country," and then elaborated on how she saw Canadian Niagarans as "an extension of Americans. I've always been a Canadian, been proud to be Canadian, but I've always thought of ourselves as the next thing to Americans ... different to [other Canadian] people that we know up North." She later added that she saw Canadian and Americans as both "different" and "the same," the latter facilitated by the Americanizing effect of exposure to U.S. radio and television.

Tom reversed Jill's description by describing the immediate U.S. side as "another Canada," but echoed her views about the influence of American media when he stated, "We watch the same [television] shows as them [those on the U.S. side] ... we have all the American channels. I'd say we're as much American as a Canadian can be. So when they [from the U.S. side] come over [to the Canadian side], I really don't even notice the difference."

Just as Jill had contrasted those living close to the border in Canadian Niagara with other Canadians living "up North," many others also offered observations that the Americanized character of Canadian Niagarans made them different from co-nationals from farther inland. Dual citizen Ashley wondered whether "[Niagara border residents] have more of an American identity than anywhere else in Canada," adding that upon meeting people from elsewhere she found that they "had a much different take on Canada than I would growing up near the border."

Ed shared that he didn't consider himself "completely Canadian," because he felt that his border upbringing had made him "almost American." He added that when he was asked about his citizenship while crossing the border, he was clear that he was not American, but also not a "a die-hard Canadian" because he was "so close to the border that it was kind of that grey area ... somewhere up around Moose Jaw [Saskatchewan] maybe you're like 100 per cent Canadian, I guess ... but living down here ... you're kind of influenced by both [Canada and the U.S.]." Ed also contrasted living in Toronto, where "everything was Canadian," to Niagara, where, as he put it, "I don't really feel like I live completely in Canada." Jim, who was applying for dual citizenship, claimed that "living in Fort Erie, [he didn't] feel Canadian ... it seems almost like you might as well call it Buffalo sometimes."

Jim also introduced a theme developed by others when he shared that relatives living closer to Toronto would tease him about his speech patterns, saying, "Wow, where are you from?" Several interviewees, in fact, reported being told by Canadians from outside the Niagara region that they sounded American. Avery, for example, described how "a few people from way up North ... they're like, 'You guys sound like Americans,'" and how she would always respond by saying, "'No, no, I don't sound like an American.'"

As these examples suggest, many described becoming aware that they were "Americanized" only through encounters with other Canadians. Some said that they first became conscious of their deviation from a standard Canadian speech variety, for example, when they attended university and other students made negative comments. Mary, for example, related that her sense of herself as a Canadian was unsettled: "When I went away to university, one of the first things that was said to me ... was 'Oh, are you American? Because you speak like an American'... I had never thought of that." She went on to clarify that she didn't actually "sound like people in Buffalo because they have a very distinct twang ... [but] to people from ... other parts of Canada [she] sounded American and that took [her] a bit by surprise." Her point about not having the Buffalo "twang" was an important one for her because, as she said, "As kids we'd laugh at that [Buffalo] accent." The stigma of the most proximate American speech variety also emerged when Nancy recalled being determined never to speak in the "dreadful way that people from Niagara Falls, New York, Western New York, did." She "was never going to get those hard Western New York inflections."

Dave also reported being taken aback when fellow university students commented on his allegedly "different accent," as he considered his speech to be "normal Canadian." He described how he would respond by telling them that they also sounded American. In the case of a roommate who, he said, pronounced "pasta" as "paa-sta" and "bagel" as "bay-gal," he'd retort, "I have an accent? You sound like you're from Massachusetts!" Because visiting Americans also told him that he sounded as though he was from New Jersey, however, he had resignedly concluded that distinctive linguistic markers were just "part of living on the border."

Jessie who also experienced university peers commenting on her speech reported:

> This opened my eyes big time because I made friends with people from … throughout Ontario and they just comment on how we [students from the border region] speak … they sound totally different from me and my family and they're just [saying], "You guys are so influenced by the American culture and you don't even realize it."

Jessie went on to suggest that these students seemed to her to have "more Canadian national identity," while she felt "like a mix. I don't know what I am." Experiencing this "mixture" as problematic, she was making efforts to remove her "twang" because, she explained, "I want that distinctiveness [from Americanness], I want to be recognized as a Canadian."

In these comments, interviewees described coming to the realization that their border upbringing was linked to speech patterns that were identified by co-nationals as problematically Americanized. Such everyday observations about difference in language use and their role in marking national identity can be contextualized within sociolinguistic and dialect studies in Canada/U.S. border regions, including Canadian Niagara. The Dialect Typography Project, initiated and supervised by University of Toronto linguist Jack Chambers (see http://dialect.topography.chass.utoronto.ca/), included the Golden Horseshoe area of Southern Ontario and Northwestern New York as one of its survey regions, and the data gathered documented the specific differences in pronunciation, choice of vocabulary, and other aspects of language use between American and Canadian speakers of English. Although the situation with respect to language stability and change is complex, it is sufficient to note here that although there are examples of convergence between American and Canadian English (e.g., the almost exclusively Canadian term "chesterfield" is being replaced by the American form "couch"), other Canadian terms such as "pop" and "runners/running shoes" (used by interviewees) remain strongly favoured by Canadians over their American equivalents of "soda" and "sneakers."

In fact, local press reporting on this research cited Chambers, commenting that the phrase "running shoes" was one of the "most clear-cut" cross-border differences to emerge from the study ("American Words Checked at Border: Canadians Along Border Retain Linguistic Identity," *Niagara Falls Review*, 5 May 2001, A5.) While this work emphasized the dramatic "drop-off" in the usage of the term "running shoes" between Canadian and American Niagara, Easson (1999) also noted that the use of "running shoes" had declined across Canadian regions, with residents of the immediate border regions of St. Catharines, Welland, and Niagara being "significantly more likely to say 'sneakers' than their fellow Canadians further removed from the border" (107).

Eassons' (1999) description of the Canadian Niagara's "significantly different variant preferences" from the rest of the Canadian Golden Horseshoe region is consistent with the interviewees' experiences of having co-nationals comment on their speech patterns, but it also explains why these outsiders' description of their speech as Americanized came as a surprise. As border residents, they were keenly aware of the much more significant "linguistic barrier" between those on both sides of the Niagara River (Easson 1999, 99) and, as the local press report on Chambers's work noted, "Border communities are often more vigilant about guarding their distinct speech patterns" ("American Words Checked at Border: Canadians Along Border Retain Linguistic Identity," *Niagara Falls Review*, 5 May 2001, A5). Canadian Niagara speech patterns, then, were regionally distinct in some minor ways from those of co-nationals in adjacent regions but more significantly distinct from those of U.S. residents immediately across the river.

As indicated, many interviewees shared a sense that their Niagara upbringing in some ways made them "less Canadian" or more Americanized than fellow co-nationals. Like others, Jim described encountering fellow university students from outside Canadian Niagara and thinking, "Man, they're so Canadian." Travel outside the region was also credited with producing this new awareness, as when Kevin contrasted Canadian Niagara border residents to "real" Canadians elsewhere:

> I went out east for a little bit, I went up north … and I mean, those are Canadians, you know what I mean? They watch the CBC [Canadian Broadcasting Corporation] … I didn't even know what the CBC was …

> now I do obviously, [because I] studied it in school and everything, but ... they listen to Canadian music, they watch Canadian TV, they're really Canadian.

While much of this commentary explicitly or implicitly suggested that an Americanized Canadianness was experienced as problematic, an attempt to challenge this emerged in Mary's description of how at first she was "a little offended" when fellow university students heard her accent as American, but then she decided,

> I'm from this kind of unique place ... Americanness is part of me ... though I don't see myself as an American, I think I have the advantage of having seen Americans maybe more than a lot of other Canadians who couldn't get American TV in parts of Canada ... I thought of them [non-border Canadians] as being less cosmopolitan. I thought they were a bit backward ... I thought some of these Canadians ... just don't know stuff.

Here, any suggestion of a stigmatized Americanized Canadianness was countered by the claim of a positive "cosmopolitanism." While some of the constructions of both sides of the river as "pretty much the same" or Canadian Niagarans as Americanized, offer hints of a binationalized hybridity in Canadian Niagara, this remained a minor theme in the interviews. Much more common were unambiguous constructions of nationalized difference, which illustrated Kymlicka's (2003) observation that Canadians tend to look for and highlight differences with Americans wherever possible (364).

Non-Americans

As mentioned above, constructions of Americanized Canadianness often focused on speech patterns, but consistent with the linguistic research cited above, speech patterns were also cited as evidence of the way in which Canadian border residents were nationally distinct from their U.S. neighbours. Kerrie, for example, claimed that when she was in the U.S. she found that her "accent" was "constantly made fun of" and she emphasized the "weird" degree of "difference" between the two sides of the river, given their geographic proximity. Dual citizen June, who had grown up on the U.S. side with a Canadian father, also described how she found it "aggravating" when American friends would denigrate the speech of visiting Canadians, but she also acknowledged that her own family would "laugh ... whenever my father got around his family, [because] all of a sudden he had a full-blown Canadian accent ... we used to say ... 'He is really Canadian today.'" Chuck, who lived on the U.S. side for his high school years, also described how his Canadianness was "always a brunt of a joke ... from the minute I opened my mouth a girl said, 'You're Canadian; say "about,"' adding that while this was "done with fun," he was always constructed as "foreign" despite Canada being a "next-door neighbour."

Dave was among those who emphasized cross-border nationalized difference when he stated, "It just perplexes me ... how different the two sides are." While he had come to accept that linguistic distinctiveness was part of a border identity, he also claimed that his border upbringing probably made him "more Canadian" than non-borderlander co-nationals. Contradicting any suggestion of a diminished Canadianness, he commented:

> I almost feel more Canadian living on the border. Because, I mean, I met so many people this year [at university] that are from farther inside of Ontario. And I know they think they're Canadian, but really, to me, Canadian is being aware of other cultures and ... embracing them.

For Dave, then – in a twist on Marys' suggestion of a link between Americanization and cosmopolitanism – Canadian borderlander exposure to and tolerance of Americanness instead made him "more Canadian." Critical to this claim was the positing of the border as a marker of Canadian-American difference with the corollary that Canadian borderlanders, through their engagement with this difference, exemplified alleged national Canadian traits of awareness and tolerance of cultural diversity. Fellow students who had less exposure to U.S. citizens than he did, Dave claimed, were less likely to embody Canadianness precisely because they were more likely to "put them [Americans] all in one group." The suggestion that exposure to the cultural difference of Americanness made Canadian border residents more aware and tolerant and therefore more Canadian than other Canadians living inland was, however, countered by others, who made the opposite argument, notably that proximity to the U.S. produced greater anti-Americanism among Canadian border residents than among other Canadians. I will discuss this in more detail shortly, but first it is important to look more closely at how the more prevalent constructions of nationalized difference on both sides of the border articulated with constructions of classed and racialized difference and inequality. I turn to these now, as they are important not only to constructions of national identity but also to inclusions and exclusions within Canadian Niagara and beyond.

Classed and Racialized Bordering

The way in which claims of nationalized difference articulated in significant ways with claims about classed and racialized differences, was apparent in interviewee descriptions of a more prosperous Canadian side, that inverted the wider asymmetry of Canada/U.S. bilateralism.[2] Michael, for example, recalled that when he walked across the Whirlpool Rapids Bridge, the evident poverty on the other side led him to conclude that "Canada [had] the advantage over the States, and you can visually see that upon crossing the river, right off the bat."

A nationalizing of alleged economic differences was also apparent in Barb's statement that "going over to the United States and seeing the poorer areas" made her "very proud to be a Canadian." As evidence of the distinct "cultures" of Niagara Falls, Ontario, and Niagara Falls, New York, Dave cited the poverty of the immediate other side, noting that "as soon as you go across that bridge ... you know you're in the United States, because the shops are boarded up ... It's a very different feeling ... you go over and it's like you're in a ghetto almost." Nancy, who was able to recall the 1950s and 1960s – when Niagara Falls, New York, was "a buzzing, booming place" – described the U.S. side as a "ghost town" at the time of her 2001 interview. Perceptions of greater poverty on the U.S. side, then, produced a sense of relative economic advantage for those in Canadian Niagara.

Miles made a point of distinguishing the immediate other side from the United States as a whole when he commented that "the United States itself can be a nice country, it's just that particular city [Niagara Falls, New York] isn't great; they have a really high crime rate and there's a lot of poverty." Others also focused on more local than nationalized cross-border disparities; Peter, for instance, remarked that "every time you're over there [other side of the river], there's always people in the parking lot ... asking for money, and we always used to find that strange because you didn't see that over here." Sophie, who would head "for the [U.S.-side] mall and then [get] out of there" because "it was a lot dirtier and less safe," shared that she was "glad that [she] grew up on this side," because "just looking at the other side ... they have no money. You can see the clear difference."

The role of class was also apparent in some of the apparently contradictory accounts of interviewees. Yvonne, who had relatives on the other side of the river, for example, suggested in one part of the interview that she did not think of the immediate other side as part of the United States – because "they're just so close to us that they're kind of a distant part of Canada." Later, however, she was discussing the "scary" experience of being in a "lower-class area" on the American side, mentioning that her parents "would lock the [car] doors ... [because] being in a different country, you never know what could happen." Here, the otherness of poverty seemed to transform the construction of the other side from an extension of the nation (i.e., a "distant Canada") to a negatively experienced "different country."

While not as frequent as references to cross-border class disparities, there were enough comments about race to suggest that this was also significant to white Canadian interviewee constructions of cross-border difference. The association of Americanness with blackness emerged in Nancy's reflections on the shopping and recreational spaces that she frequented on the other side in the 1950s, to which she added that on the U.S. side "there were black people, unlike the Canadian side of the Niagara border," where she had grown up in a "very white community." Dale also recalled the "stark" contrast between the two sides of the border and thinking, as a child crossing into the U.S., "Wow, there are a lot of black people who live here."

Blackness could be intertwined with poverty and danger in white Canadian border resident constructions of cross-border difference. Mary, who described the Canadian Niagara of the 1960s as having "very few ethnic minorities," recalled how U.S. "race riots" led her peer group to shift from crossing at the Whirlpool Rapids Bridge, where the other side was poorer and "more dangerous," to the allegedly safer Rainbow Bridge. Rather than walk over, she also noted, "You would drive and ... park as close to the place [you were going] as you could." Buffalo, she added "was even more scary in that way because there were very large ... black areas."

Ed recalled his childhood awareness of how some areas of Niagara Falls, New York, and Buffalo were "segregated": "The black people seem to live with black people, and the white people seem to live with white people ... [and in] run-down neighbourhoods ... there'd be a lot of black people." When he was a young teen, he recalled, his family's car broke down in Buffalo and they walked to a garage through an area where "everyone was black"; he recalled thinking for the first time, "'This feels weird'... being a white person in an all-black neighbourhood. I was like, 'This doesn't feel right. I feel out of place.'" He contrasted this to the Canadian side, where "everyone's mostly white" and those of "a different ethnic group [are more] scattered about."

Katherine, reflecting on the 1980s and 1990s, reproduced a similar discourse:

> It was sort of scary ... especially if you go into downtown Buffalo. It's all slums and everything ... I remember my parents brought us back one night and ... we drove through Buffalo, and it was just so scary because it's all, like, run down and everything, and you're just like, "Wow, this is so different," because around here you don't see that much of it at all ... and they have a large coloured population there too, which I'm not used to around here. Like, it's really white ... there wasn't a lot of that at all in my school. So that was really different.

Melissa also contrasted her Canadian-side "middle-class" neighbourhood with the "really seedy neighbourhood" in Buffalo where her family bought her some tap shoes, recalling the latter as being "like a different country ... dirty and ... really scary." The presence of "more black people" there, she added, led her to "being a little intimidated" and feeling "like I had a Canadian stamp on my forehead."

As with class, apparently contradictory constructions of the other side illuminated the significance of race. Susan, in part of her interview, emphasized cross-border similarities when she described how those on the other side "look the same, they talk the same, it's the same TV, everything's the same, except, you know, a little bit of an accent." Elsewhere in the interview, however, she emphasized cross-border differences, including the observation that on the Canadian side "everyone" was white while the American side was "white and black." Jim, who had discussed the lack of any "culture shock" when crossing the border, also noted that the presence of "black people" in Buffalo contrasted with his "very white, Anglo-Saxon Protestant" hometown of Fort Erie. Cross-border difference constructed in terms of class and race posited a Canadian Niagara marked by relative prosperity and (sometimes, but not always, explicitly named) whiteness in ways that obscured both historical and contemporary regional realities of classed and racialized diversity and inequality. Such local constructions of nationalized difference were furthermore, congruent with dominant white-settler Canadian nationalism, which positions Indigenous and ethnoracialized Others as peripheral to, or completely outside, imagined Canadianness.

A striking aspect of the construction of a white Canadian Niagara was the way in which this also appeared in other contrasts drawn between inland Canadians and Canadian Niagarans. In particular, there were references to a more multicultural/multiracial Canadianness to be found closer to and in the Greater Toronto Area. Anna, who had spent some time at a Toronto university, for example, first described Torontonians as "a thousand more times accepting than people in Niagara … of race, of anything." She then went on to describe the "very multicultural" university that she attended there as "very politically correct," adding that if you were not politically correct you could "get in a lot of trouble" and that she "almost felt like a minority because there are so many different people there."

Likewise, Sophie contrasted the "different neighbourhoods and different ethnicities" of Toronto to Canadian Niagara, which was "so white … bland and just really small town." She first described Toronto as "really nice," but went on to offer a more ambivalent account of how a problematically open Canada border seemed "to let most people in," as evidenced by the fact that "if you look at Toronto and Mississauga … it's mostly ethnic now." Such characterization echoes O'Connell's (2010, 542–3) observation that constructions of Canadian urban spaces as "multicultural" and less urbanized spaces as "non-multicultural" can serve to naturalize the allegedly white spaces of the latter by obscuring not only the ongoing presence of Indigenous and ethnoracial communities but also the histories of colonial and racialized dispossession and exclusion that produce these allegedly white, non-urban spaces. Contrasts that were drawn between a putatively white Canadian Niagara and the U.S. side of the river on the one hand and Canadian cities of Toronto and Mississauga on the other, hinted at other constructions, discussed in chapter 6, of a white Canadianness under threat from ethnoracialized, globalized Others. Such everyday nationalism in the border region, then, reflected and reinforced, often intersecting ways, classed and racialized inequalities in the borderland and beyond.

U.S. Visitors and Anti-Americanism

Perceptions of those on the other side of the Niagara River as well as co-nationals further inland shared in the interview setting provided important insights into varied constructions of everyday borderline

Canadianness. Interviewee commentary on American visitors to Canadian Niagara was also marked by strongly nationalized discourses. While Canadian Niagara attracts a wide variety of global tourists (and I analyse commentary on non-American visitors in the next chapter), the focus of Canadian border interviewee discussion of tourism in Canadian Niagara was on the U.S. citizens who made up almost 40 per cent of overnight visitors in 2004. In stark contrast to official cross-border regionalism, which aimed to cultivate greater cooperation around cross-border tourism, interviewee discussion of U.S. tourists contained the clearest examples of nationalized identity of Canadianness being defined in opposition to a negatively constructed Americanness. The interview material that I gathered represented an updated version of a longer history of local ambivalence about U.S. visitors to Canadian Niagara Falls (Dubinsky 1999, 210).

As mentioned previously, some interviewees linked border life to greater tolerance of Americanness, but it was also suggested that frequent exposure to Americans could intensify anti-Americanism. It was the latter point that was most commonly made in the context of often highly charged discussions of U.S. visitors, as many suggested that there was a link between negative experiences with such visitors and the development of antipathy to Americans.

The broader observation that expressions of Canadian identity are often linked to anti-Americanism is well documented (Granatstein 1996), and Mackey (1999) has made the point that "while one of Canada's defining and supposedly essential characteristics is tolerance to difference, one of the major socially acceptable forms of intolerance has been that directed at the United States" (145). That anti-Americanism was a feature of border childhood in Canadian Niagara was suggested by Richard's story about studying the War of 1812 in elementary school; when his class was told that "we had won the war and we sent them [Americans] ... packing ... everyone in the class was like, 'Yeah, yeah, we beat America!'"[3] Many, however, suggested that their own and other border residents' anti-Americanism had increased with age, and they offered the additional suggestion that this was the result of negative experiences while serving U.S. visitors as youthful employees in Canadian tourism-related jobs.

Interviewee commentary on American tourists was primarily nationalized but was also articulated with classed, racialized, gendered, and regionalized constructions in significant ways. There was, for example, deeply classed discourse about the wealthy U.S.

visitors with Canadian-side vacation homes mentioned earlier. Some interviewees had seasonal jobs serving these well-off cottagers, described by Kevin as "upper-class white people … doctors and lawyers," whose "cottages" were better described as "mansions." Lisa, who taught swimming to some of the children of these Americans, described them as "upper class" but distinguished between the "very upper class … old rich money," who were "the sweetest people in the world," and the "new rich," who were "snobby" and had "brattier" offspring. While some of her Canadian co-workers, she suggested, portrayed Americans as "stubborn" and "rude" and had trouble with the "pushy" parents, she emphasized that she had found the U.S. parents of those in her swimming classes to be "very appreciative" and financially generous. When she and her co-workers organized a beach party for them, she noted, "the amount of money we made was unbelievable."

Like Lisa, Kerrie, whose family rented cottagers to Americans, contrasted her more positive childhood views of these renters with what she saw as a rampant anti-Americanism among her Canadian friends. She noted that her Canadian friends had directly challenged her pro-American views, and in response to this peer pressure, she had adopted a more anti-American stance as she grew older:

> I wanted to be an American for a long time … [but] my [Canadian] friends … would be, like, "No, you don't want to be an American." And I was, like, "Yeah, I do." They're, like, "You're not our friend"… they would be, like, "I hate Americans. They're rude, they're this" . . . [but] I loved them [Americans] because everybody that rents [cottages] off of us is so nice to me … nobody was ever rude to me … But … [my Canadian friends said], "How could you want to be American?"

Dual citizen Kevin described developing friendship and dating relationships with American summer visitors but also described the "flip side" of this in the form of considerable tension between Canadian and American youth. He described those at his school as being

> really strong Canadian and anti-American … You'd be at a beach party … and people would be drinking, and … there would be fights between Americans and Canadians. Like, "Get out of our country … this is our country, you guys invade it in the summer." So, as much as there were friendships, you also saw that side of stuff.

Joe also used the terminology of "invasion":

> Some of my friends … think that the Americans are … starting to invade us … This past summer we were going to a beach and we … tried to cut across some properties and houses, and this guard came and said, "Hey, come over here." And we had to go around … my one friend was really resentful that … the Americans here were taking our land.

Other U.S. visitors who attracted negative commentary included young people described as frequenting Canadian-side clubs. Dan, for example, indicated that, like other locals, he had stopped going to nearby venues because, in his view, a lower Canadian drinking age attracted young people with weapons, whose American "over-patriotism" resulted in violence. While noting that avoidance of bars frequented by U.S. visitors might be an example of "stereotyping," he added more ambiguously, "I don't know if that's racist or whatever, but obviously they have a reputation." Yvonne also linked U.S. nationals to violence in Canadian Niagara, while Amanda, who also avoided bars in downtown Niagara Falls, Ontario, "because … there's so many Americans there," added that a friend of a friend who worked as bouncer reported that Americans would say, "What do you mean I can't have my gun? Or what do you mean I can't have my switchblade with me at the bar?" She also linked "African-Americans" in particular to guns and "trouble."

Ed, who claimed that he could "pick an American out of a crowd," because they were marked by distinctively gendered and racialized hair and clothing styles, went on to describe the phenomenon of intoxicated young males "squaring off on the street" with nationalized provocations. One from Western New York might say, "I'm an American, what are you going to do about it? … I've been in the Marine Corps," to which a young man from Toronto might aggressively counter, "I don't care. So what you were in the Marine Corps … let's see what you got." While portraying visitors from the U.S. as the instigators of violence, Ed claimed that he personally related to those who came over to have "a good time," just as he used to do "on the other side." Ed was among those who claimed that his border upbringing had produced greater tolerance of Americans and that other Canadian border youth as well were less likely (than the hypothetical Torontonian of his story above) to become imbricated in such nationalized physical altercations.

In contrast to other interviewees who expressed resentment at being repeatedly asked for directions to Canadian-side strip clubs by U.S.

tourists, Kevin's comments were more positive about at least some of these visitors commenting, for example, that it was "cool" to see celebrity male athletes from the U.S. side in these "facilities":

> We would be eighteen, we would have our fake IDs, and we would go into the strip clubs ... And [would see] ... players from the [Buffalo] Bills [football team] in there and from the [Buffalo] Sabres [hockey team]. And we thought that that was just really cool that they were in our town partying it up.

More common than this kind of positive remark, however, was the suggestion that encounters with U.S. visitors produced greater intolerance among Canadian border residents. Avery was one of those who speculated on a link between border life and the production of anti-Americanism when she reported, "It just feels like it's inside you that you're supposed to hate Americans. I don't know if that has to do with living on the border, but I think it's just because you're confronted with it [American presence] every day." Debbie also suggested that anti-Americanism was characteristic of locals because "growing up we are just so proud that we are on this side of the border."

Both possibilities – that border life could result in both greater tolerance and intolerance – appeared to be considered by Anna, who claimed that "[living at the border], you've got more of a first-hand experience ... you always hear jokes about the Americans, [but] ... it's kind of different when you actually see ... and ... talk to them. " At the same time, she added that her friends learned anti-Americanism from their parents and that "people on borders ... are so proud of where they live ... that they aren't accepting of anyone else." Jill pointed to Canadian border residents engaging in anti-American "slurs," saying things like "Oh, those American drivers" and "All these Americans over here," while Tom acknowledged there were "two sides of the coin" when it came to border living and attitudes towards Americans because

> some people who ... live in the border town ... are going to go, "You know what, we're [Canadians and Americans] all the same, we're all alike." But ... there are [also] some people who think Americans are jerks ... You get cut off [by an American driver and say], "Oh you Americans, so typical of you," and stuff like that.

The paradoxical combination of local complaints about U.S. nationals and frequent crossings to the U.S. side to benefit from lower gas prices was highlighted by Zak, who noted, "Growing up, we were like, 'Oh, those Americans!' But then, you go over there for gas, we're like, 'Yes, yes, yes.'"

As indicated, many specifically linked teenage work experiences to the production and/or intensification of anti-Americanism, while acknowledging the importance of American visitors to the regional economy. Yvonne (who had relatives in the U.S.), for example, recounted that she "would complain when an American came in [to the restaurant where she worked as a server], but ... if they hadn't come in, then we wouldn't have a job ... I wouldn't be making nearly the money that I did in tips."

That the Canadian Niagara tourist industry seemed to "bend over backwards" to cater to U.S. visitors was problematized by Michael, but he then went on to add that when his Canadian friends critiqued U.S. tourists he felt like reminding them, "You have some of your jobs because so many of them are coming over and it helps our economy here ... it's a give and take, so if you could just put up with it ... you could benefit." He noted, however, that among his acquaintances "distaste for Americans" was common, and he attributed this to American visitors' lack of knowledge about Canada and their attitude of "superiority."

Many couched anti-American comments in ways that signalled concern about being perceived as stereotyping or being racist in generalizing about U.S. visitors. Some, for example, prefaced their negative commentary with a disclaimer, as in the case of Debbie who, in speaking of her work at an inn, noted, "We have a lot of Americans come in ... I'm not racist, but I don't like dealing with them." Others presented themselves as having more nuanced attitudes than those around them. Emma, for example, suggested that many Canadian-side tourism workers saw Americans as "a little rough around the edges," "impatient," and "rude," but added, "I don't know, I think you get that with every nationality." Dave, who presented border residents as more tolerant of Americans than non-border residents, confided, "Sometimes I catch myself ... thinking Americans are always rotten, but ... I mean, no matter where they're from, you're always meeting people that are jerks."

Miles also worked to counter any perception of undue anti-Americanism by focusing instead the challenges of serving tourists as a whole, not just American tourists:

> I think a lot of people regardless of where you're from, when you go on vacation, people tend sometimes to be a little bit more rude or demanding

> ... a lot of the Americans that come in are on vacation, and sometimes you have to realize that, "Okay, they're on vacation, they're a little bit stressed."

Barb also suggested that her own negative views of American visitors might have stemmed less from any inherent characteristics of U.S. tourists than from working in a service industry that she described as

> a tough employment area ... [where] you are typically not treated very kindly ... especially during the summer ...[when you have] tourists coming in and being ... not exactly the nicest people to you ... I was tired of being yelled at for things that were not my fault.

Hannah, too, worked to distance herself from any suggestion of anti-American stereotyping by distinguishing between different work sites noting that when employed at a bed and breakfast, she got to know tourists from the U.S. and found them "friendly," but when she entered a retail job she "got a different opinion ... [because American customers] were rude and demanding." Dave's story about an "exceptional" American who said, "I love your country, you guys are the best," while also perhaps intended to demonstrate that he was not anti-American, however, actually served to reinforce his otherwise negative portrayal of U.S. retail customers.

Another way in which some appeared to deflect an anticipated critique of indiscriminate anti-Americanism was by emphasizing alleged regional distinctions among U.S. tourists. Jason, for example, suggested that those living on the immediate other side of the river had "more awareness of Canadians and Canadian identity than people that live further in," while others specified that their greatest animosity was directed at the most proximate Americans (e.g., those from Buffalo, Tonawanda, Western New York, New York, or the northern states).

Yvonne was among those who offered particularly negative portrayals of the Americans living on the immediate other side of the border when she explained that, while she and her fellow restaurant workers considered U.S. visitors to be "cheap," this was not true of "all Americans" just "the ones right over the border," who "never tipped." Likewise, Amanda (who had U.S. relatives) said, "I couldn't stand talking to the people from Tonawanda, but I could talk to people who were from the Southern states for hours. They were so nice and they were so pleasant. But the people from Tonawanda ... they ... had that egotistical ignorance." Barb also related that fellow workers perceived those from the Southern states to be "very nice, very pleasant, easy going, good people to get along with," in contrast to those from neighbouring New York State, who were considered to be "more arrogant, rude, straightforward, to the point, and not so … tolerant if somebody were to make a mistake."

U.S. visitors from the immediate other side (with the exception of wealthy cottagers), were more likely to be same-day cross-border shoppers or the clubbers already discussed. That distinctions drawn between short- versus longer-distance U.S. nationals could also be classed and racially coded occasionally became explicit, as when Dale described most U.S. visitors to Canadian Niagara as "working-class … the ones who can't afford to fly to Europe … they drive from Cleveland to Niagara Falls. It's the family vacation," and dual citizen Rob's suggestion that the U.S. visitors to Canadian casinos were "minorities," who "basically blow their money and then the States has to support them afterwards."

Just as visitors from the U.S. were regionalized, classed, and racialized, they were also gendered, as has been documented in other tourist and border settings.[4] I mentioned earlier how the "other side" could be constructed as particularly dangerous to young female Canadians. When it came to U.S. visitors to Canadian Niagara, gendered and sexualized constructions sometimes also came to the fore. Kerrie, for example, offered the comment that visiting U.S. males were too "forward," while Sheila and Yvonne (despite their own kinship ties to the U.S. side of the border) shared negative views of cross-border intimate relationships. Sheila claimed that "the American men … [visiting a club in Niagara Falls, Ontario] have no shame; they'll just grab you … we have a rule … don't date American boys." Yvonne related that her father told her, "You'd better not bring home an American," and suggested that there was little cross-border dating or intermarriage in the border region, due to widespread negative views of those on the U.S. side. As she put it, "It's a big stereotype … but the people over there? We just … don't associate with them." Anna, meanwhile, when describing a U.S. national engaged to a Canadian friend, said that while it was "cool" to meet people of different nationalities and the American fiancé was "pretty nice," she was suspicious of his motives "because he's from the States."

Despite references to Canadian women in cross-border intimate relationships with men from "over the river" – such as Lisa's friend of a friend, who crossed "back and forth" while dating, and Barb's reference to the possibility of meeting American men at the summer Friendship Festival – gendered (and perhaps racialized) regulation of young white Canadian women may have created greater barriers to U.S. male/Canadian female than Canadian male/U.S. female relationships. This is suggested by Zak's reference to how his male friends would seek out and enjoy relationships with girls from the U.S., whom he contrasted with more conservative "farm" girls on the Canadian side. Constructing U.S. women as more sexually available, he noted that his friends would go to the "all-ages club ... where all the Americans come ... for us that was the appeal ... [there were] new people and we could kiss them, we could actually do things." A gendered patterning of Canadian/U.S. courtship was further suggested by the fact that four of the five dual citizens interviewed had U.S. mothers, and of the three interviewees with U.S. or dual Canadian/U.S. citizen parents, two had U.S. national grandmothers.[5]

Expressions of anti-Americanism, then, were articulated with classed, racialized, regionalized, and gendered borderings in complex ways. Such complexity, however, did not preclude unequivocal assertions that all U.S. visitors shared collectively nationalized traits that allegedly made them more difficult to deal with than other international visitors.

Dual citizen Ashley, who had worked at a major attraction in Niagara Falls, for example, stated, "We always had the stereotypical joke about the pushy American customer and ... the bad thing was, it was true." Working in tourism, Anna noted, meant encountering "a lot of different cultures," but while "you deal with people from everywhere ... [Americans] are the most ignorant and the most disrespectful." Susan claimed that those who worked at the Niagara Parks Commission "really start to hate Americans after a while," and Dale stated, "To grow up in Niagara Falls means that you worked in the tourism service sector, which I think contributed to your hatred of Americans ... surprisingly, not other cultures, it was just Americans."

According to many of the accounts that linked tourist-related employment to anti-Americanism, a central source of frustration revolved around money. An intensely narrated subtheme was that of monetary friction resulting from U.S. visitors' alleged tendency to contest exchange rates and to resist receiving Canadian currency as change. Hannah explained that in her workplace setting, "it would be very stressful ... you'd post a sign on the door ... saying what the dollar is at and it didn't matter. They'd [American visitors] still want to change it [the posted exchange rate]."

Jessie commented that American visitors would get "ticked off" if they did not get U.S. currency back and were "always asking, 'If I give you American, will you give me American back?'" Dual citizen Ashley recalled that while working at a major tourist attraction, "we would always have Americans very angry ... that you couldn't give them their own money back ... I would always think, 'Did they not think about this before they came over [the river]?'" Brenda described how, as a worker, "you would get yelled at when you gave Canadian money back" and how U.S. visitors would deride Canadian currency as "monopoly money ... and they'd get mad." She described her frustration with having to repeatedly tell them that they could spend Canadian currency anywhere in Canada and noted that she wanted to add, 'You don't want it? I can find a use for it here.'"

Many echoed this annoyance with American depictions of Canadian currency as "monopoly," "play," or "funny" money. Amanda related her visceral reaction to being approached by U.S. customers because, she claimed, she knew she would "have to sit there and listen to them say things like 'I don't know what this funny toonie loonie [Canadian nickname for two- and one-dollar coins] thing is.'" She added that she knew that she would have to "argue with them about their change ... and why prices [weren't] in American" and claimed, "That's from living around here and growing up around here." She expressed regret at her increased tendency to stereotype Americans as a result (as she saw it) of her work experiences.[6]

While some alluded to the monetary advantages of American currency in the context of the lower Canadian dollar of the late 1990s and early 2000s – Joe for example, described his appreciation for U.S. money as a "world currency" like "gold" that was "worth more," and dual citizen Dylan favoured Canadian adoption of the U.S. dollar – many more highlighted their resentment about a perceived lack of monetary reciprocity. Sheila complained that in New York State "they won't even accept a Canadian penny, whereas we'll take anything that's American," while Amanda added "[Americans] think they can just cross the border and use their money ... but we can't do it to them ... they won't even take Canadian pennies."

More significant than this perceived asymmetry, however, was the suggestion that the reported monetary behaviours of U.S. tourists indicated a broader lack of appreciation of the significance of having crossed an international border. Interviewees portrayed the reluctance of U.S. visitors to accept Canadian currency as evidence of an allegedly nationalized tendency to disregard, or not comprehend, the reality of Canadian national space and sovereignty, rather than a pragmatic tourist economic strategy (e.g., resisting unfavourable exchange rates and attempting to reduce double exchange processes).

Amanda suggested that this was particularly true of those from Tonawanda:

> [They] think the world should revolve around them [and that] all of the prices should be in American [on the Canadian side of the border], all of the change should be in American [currency], and there's no respect for our "funny money" as they call it ... And it drives me insane when people say that, you know: "Are your prices in American?" "Well no, you're in Canada, you're not in the States anymore ... you go anywhere else in the world and you exchange your money."

Sheila also described her frustration with U.S. visitors asking whether prices were "in American," explaining how she wanted to say in reply, "You must have been sleeping at the border; didn't they ask for your citizenship?" (A comment that problematizes not only American border crossers but also a Canadian border that apparently does not "wake them up.") She suggested that those of "a different ethnic background," in particular, "didn't understand the concept that they were in a different country," and how she felt like telling them that "the bills are different because it's a whole different country. Just because you crossed the border and it took five minutes, doesn't mean that you aren't somewhere else." U.S. visitors (and, as some of these comments suggest, perhaps especially racialized minority U.S. visitors), then, were critiqued for allegedly not respecting an ideal congruence between national territory and currency, while simultaneously faulted for precisely such congruence in allegedly not accepting Canadian currency in U.S. territory.

Stories of monetarized conflict point to what Zelizer (1998) calls "the intimate interplay between monetary transactions and the construction of social relations and meaning systems" (1376). Donnan and Wilson (1999) note of cross-border shopping more specifically that such exchanges go beyond interpersonal exchange to "group economic, political and social relations" (119).

Stories about money certainly reproduced claims about allegedly nationalized U.S. arrogance and/or ignorance vis-à-vis Canada and Canadians.[7] "[American visitors] would insist on getting American change and not even realize they were in another country," said Dale.

"We just thought that they were brain dead or something." As Miles put it, "Americans would come over" and then ask "how much farther to Canada?" apparently not realizing or acknowledging that they had crossed an international border. Such behaviours, again, were described as particularly galling when displayed by the most proximate Americans, whom Tim described as "clueless about us ... even though they live right on the border." Dave echoed this claim, suggesting:

> [Americans on the other side of the river] haven't got a clue about Canada. I went to class with a guy from Buffalo, he didn't have a clue ... People from Niagara Falls, New York, or Tonawanda or whatever, they come over and they wouldn't have a freaking clue. They wouldn't know that the capital is Ottawa, that type of thing ... they wouldn't know the name of the prime minister.

U.S. visitor ignorance of things Canadian was also invoked in stories meant to illustrate another alleged national attribute of "stupidity." Many stories thus emphasized the "stupid questions" asked by U.S. tourists. Ed, for example, described Americans wondering what country and/or time zone they were in and "what time they [had] to be out of Canada." American stupidity was also supposedly illustrated in stories about U.S. tourists having inflated or inaccurate ideas about national differences. Richard, for example, recounted how he and his friends were "insulted" when a visitor from Buffalo, who had "no idea what was on this side of the border" approached them, asking, "Hey, y'all speak English?" while Michael was annoyed that some U.S. visitors assumed that those on the Canadian side were all fluent in French. In the latter case, the U.S. visitors' apparent familiarity with Canada's official bilingualism but lack of awareness with the reality of low rates of bilingualism in Canadian Niagara was portrayed as further evidence of U.S. ignorance.[8]

The most apocryphal of stories about U.S. visitors, however, related to those from the U.S. allegedly expecting wintry conditions on the other side of the border. Susan recounted an incident that occurred when she was working at fast-food outlet:

> An older couple [Americans from Texas] come in, and they were questioning me as to why it was so warm. It was July, and they said they wanted to see penguins and polar bears, [saying] "This is Canada isn't it? ... Why is it so warm?" ... And I told them ... "Well, I'm afraid if you want to see

> polar bears, you're going to have to go to the Toronto Zoo or the Buffalo Zoo"... I don't think they were kidding. They really did seem surprised that it was so warm ... it was the first time I actually saw a real person really believe that, and that's when I started questioning, you know: "What do you [Americans] really think of Canada ... What do you think we are here?"

While this account was told as a first-person experience, others, like Katherine, referenced stories told by others about how Americans "come to the border and they're, like, 'Where's the snow?'" Richard described a tourist from the U.S. asking his boss if there was a heat wave, and when his boss replied that the warm summer weather was normal, the visitor allegedly accused him of lying, saying "This is Canada. There has to be snow here all the time." His friends working in tourism, said Dan, all "think that Americans are a little bit slow ... they come over in July with skis on their roof and ask about going skiing and things like that ... we always kind of joked around about that." Michael described how U.S. summer visitors arrived "expecting to be able to ski right as soon as they cross the border," and Erin complained that "a lot of Americans are really ignorant, like in terms of 'We're in Canada and there's no igloos'... Stuff like that."

Avery found U.S. visitor questions like "Where's all the snow up there, polar bears, and stuff like that?" surprising, given that Canada was "right across the border," and she felt that while "we have to learn all about Americans ... Americans don't have to learn about Canada, so you just kind of hate them for it." Nancy explained that while she expected little from European visitors, Americans being "so ignorant, so insular about, Canadian culture" is what "annoyed you. Were we not important enough?" Miles also thought it "kind of strange that we would know so much about them and they wouldn't really know much about us."

Dale also ascribed a friend's "hatred of Americans" to their "stupidity," and the latter was reinforced through the circulation of stories about U.S. visitors' "ridiculous" questions. As Miles described it:

> My dad [told] some stories about people coming across the border just asking really ridiculous questions. And sometimes I'd half believe him ... but then I started hearing the questions myself, and I was thinking, "Oh my God, people actually say these things." Like, I actually heard, "When does the Falls turn off?" from one person.

Reflecting on the experiences of young tourist workers, Sheila also mentioned how "the same jokes [about visiting Americans] go round and round, like, 'When's the Falls shut off?' 'Oh, we roll up the sidewalks and deflate the trees.'" Erin also referred to being asked "dumb questions like, 'When do they shut off the Falls?'" while Brenda added, "If people told me some of the questions that they said they had heard and I hadn't heard them myself, I would never have believed it … like, 'When do they roll up the sidewalks? When do they turn off the Falls? … Where are the skis?'"

The 2005 Pew Global Attitudes Project documented an increase in broader Canadian anti-Americanism during the period when these interviews were conducted, noting that "Canadians have a generally more negative view of American character traits than do the publics of other traditional U.S. allies" (Pew Research Center 2005, 15).

Anti-Americanism was, in fact, a potent political issue in the immediate post-9/11 period as well as at the 2003 U.S. launch of the War on Iraq, as right-wing opposition politicians decried the "anti-Americanism" of the governing Liberals. In Niagara, a visiting Stephen Harper, then leader of the Canadian Alliance Party (and later Conservative prime minister), attacked the Liberals for "pandering to rank anti-Americanism … [that] … does not serve the interests of the Niagara peninsula" ("Relationship with US Key Political Issue," *Niagara Falls Review*, 13 September 2003, A2).

The issue of anti-Americanism and tourism was also raised in the context of Ontario's government-sponsored Travel Intentions Studies, which began in 2003 with the goal of addressing the decline in numbers of U.S. visitors. A June 2004 report, for example, noted that the barriers to visiting Canada (referenced by surveyed Americans who mentioned any issues at all) were, in descending order of importance, the outbreak of SARS (severe acute respiratory syndrome), difficulties or delays at the border, terrorists (in Canada), and anti-Americanism (Ennamorato 2004, 33). Anti-Americanism reappeared in the subsequent spring 2005 report as the top issue, prompting the author to comment, "There is growing sentiment in the U.S. that anti-Americanism is on the rise in Canada" and to suggest that this sentiment might in turn be "the main impediment to growth in the U.S. source markets [for tourism to Ontario]" (Ennamorato 2005, 77). By the summer of 2005, anti-Americanism was listed second (after border-crossing concerns), in a finding that received wider attention, including mention in *The Economist* ("The Unfriendly Border," *The Economist*, 27 August, 2005).

Stronza (2001) has pointed out with regard to other tourist settings that local residents may use humour "to ridicule tourists" in ways that allow them feel "empowered by interactions with outsiders" and "to redefine who they are and what aspects of their identity they wish to highlight or downplay" (272–3). Here, stories about tourism-related employment constructed a symbolic boundary between a negative American Other and individual and collectivized Canadian Self. Anna recounted, "Dealing with stupid questions [from U.S. visitors] makes your ego go up a little bit more ... I have a lot more [Canadian] patriotism than ... when I was younger." Brenda commented further that her work serving U.S. tourists made her "realize that we are not all the same ... we definitely do have different attitudes," because Canadians, she felt, had "more tolerance ... [were] not so full of ourselves [and were] more open to other people and other ways."

Along with the Canadian border residents' humorous and less humorous everyday expressions of anti-Americanism directed against U.S. visitors, there was some broader commentary "directed at Washington," as Zak put it. Anna, for example, a month after 9/11, described how the omission of Canada from U.S. President Bush's list of countries thanked for their assistance on 9/11, made her "a little bit more hostile [towards Americans]." There were other references to U.S. policies. For example, Katherine mentioned her opposition to the U.S. death penalty and, like many, referenced a better Canadian health-care system. Her own family's "low income" position, she felt, made her attuned to the plight of those living in poverty in Buffalo because U.S. citizens

> have to pay for everything ... [and] if you just think of Americans as big built-up cities like Atlanta or something, you say, "Well, they've got money; they can afford it." But, when I ... see the slums in Buffalo ... you say, "How can those people afford to, like if they have cancer or they have to go see the doctor on a regular basis?"... So you see how our [Canadian] system is better than theirs.

Chuck also described how during his sojourn in the U.S., the cost of health care was one of his "biggest worries," and he told his friends if he was "ever seriously injured" to bring him back to Canada. June, who had grown up in the United States, noted her preference for the Canadian health-care system over a U.S. system that she felt left her parents "struggling with paperwork galore." Ed also preferred the Canadian system to the U.S. "class-based health care system where if you have more money ... you get better health care."

Other comments pointed to a Canadian public education system that, it was suggested, offered better quality schooling for all, not just those with "means." Amanda linked the importance of the border to a Canadian identity based on "our health-care system and ... education the way it is." Mary, who felt that Canadians had a better "sense of community" regarding health and education, also suggested that nationalized differences were apparent in bilateral negotiations over Peace Bridge expansion. She was critical of a U.S.-proposed new "signature" bridge" that would "destroy the neighbourhoods on both ends of the bridge," adding that the U.S. should instead spend the money on social programs to address its "terrible social problems." Avery stated that while Americans seemed to think that, being more powerful than Canadians, they were "hot stuff," and she felt that in contrast to the U.S., "in Canada, we got health care and stuff ... it's a better place to live ... it's safer. They... [have] more... guns and ... people dying all the time. You don't hear about people in Canada ... getting shot in the alley." Such claims paralleled a dominant Canadian nationalism of the time that positioned Canada as a "kinder and gentler country" than the U.S. (Kymlicka 2003, 364).

As mentioned, however, most attributed their anti-Americanism not to political or policy differences but to everyday encounters in the context of tourism-linked employment. Among other effects, such intensely narrated stories about U.S. visitors deflected attention from the many other challenges associated with precarious youth employment in a seasonal tourist service sector that had not replaced higher paid unionized jobs lost in the course of regional context of deindustrialization.[9]

The anti-Americanism found in these interviews diverges quite starkly from the pro-tourism boosterism of regional elites, specifically the tourist industry's Come Stay with Friends binational campaign, and suggests, at best, local ambivalence vis-à-vis projects of official cross-border regionalism. As mentioned, some interviewees did reference the then still binationally organized Friendship Festival, and Judith described cooperation between Niagara Falls, Ontario, and Niagara Falls, New York (what she called "both sides of the city"), in the context of a cross-border winter Festival of Lights and "emergencies" such as rescuing tourists from mishaps in the Niagara gorge, but there were few other references to cross-border regional cooperation.

American-Canadians?

As indicated, some interviewees suggested that the strength of anti-Americanism in Canadian Niagara extended to informal sanctions against the formation of intimate heterosexual relationships between Canadian and U.S. border residents. Despite alleged pressure against cross-border liaisons (perhaps especially if these were interracial in the case of young white Canadian women, as suggested earlier), the five interviewees who were dual citizens and the three whose parents were either U.S. nationals or dual Canadian/U.S. citizens were part of a wider border reality. Lisa suggested that there were a lot of "dual-citizen kids" in the region, and other U.S. relatives mentioned in interviews included Rachel's grandfather, who had been born in States; the "mixed" Canadian-American marriage of Lisa's uncle and aunt; and the relatives on the other side of the river visited regularly by Richard, Yvonne, Dave, and Amanda. I have already briefly discussed the experiences of dual Canadian/U.S. citizens, noting how their status was often invisibilized in border crossings. Here I return to this category to explore their particular binationalized positionings.

Although there were suggestions that dual citizenship could provide broader binational economic options and was sufficiently valued that some American women residing in Canada deliberately give birth on the U.S. side in an effort to ensure U.S. citizenship for their Canadian children, "American-Canadian" identities were also described as stigmatized in Canadian Niagara. This was clear in the words of a mononational Jessie, who shared her dislike of American-Canadian identities: "There was a lot of people in high school who [would say], 'Oh, I'm half American' … a friend of mine … his mother is American … I don't want to say I go, 'Eww,' but you're just like 'Oh, I wish you were all Canadian,' you know?"

Dual citizen Ashley, whose sister, in the aftermath of 9/11, had told their American mother for the first time that she was proud to be American, described her dual citizenship as an advantage because she could "work and live in either area," but also noted that she had many friends who "didn't particularly like Americans" and added, "[This was] tough for me, my mom being American." When she was younger, she recalled, her mother had been "very upset" about a "very anti-American" Canadian teacher, and she shared her mother's view that "there was a huge anti-American sentiment [in the Canadian border region]." Her mother was distressed when border Canadians engaged in "[U.S.] accent imitations ... [and] didn't particularly like the American jokes." In her own case, upon hearing an "anti-American joke," she would usually think, "It's just in good humour," but she explained that if she thought that a comment went beyond what she considered appropriate, she "might gently just say, 'well you know, my mother's American ... and I really don't appreciate anyone who dislikes Americans.'" Acknowledging that she herself would sometimes tell anti-American jokes and tease American friends about their accents, she added that she felt that she could do this "because I'm American too, even though I don't usually admit that."

Allegations of anti-Americanism were also made by dual citizen Kevin, who recounted that when he decided to attend university on the Canadian side after a period of study in Buffalo, he felt that this would require adopting a more exclusive (i.e., non-American) Canadian identity. He told himself, "I'm a Canadian. I was born in Canada. I have to lose this love I have for the States." Despite this resolution, he described a struggle to contain the "strong influences" of his "American blood" and was critical of the Canadian campus where, he claimed, "You can't tell people ... 'I'm American' or 'I'm proud to be American' or 'I have American family' [because] there's such a strong anti-American feeling."

The challenge of an American-Canadian identity was also apparent in the commentary of dual citizen Dylan, who had spent part of his childhood on the American side and continued to visit family there. On the one hand, he suggested that due to the Niagara region's mixture of Americanness and Canadianness, "if they [elite advocates of official cross-border regionalism] were to say 'We're now claiming Niagara Region International,' it would not surprise me," but on the other hand, his experiences of being called "Yankee" on the Canadian side of the border and "the Canadian" on the American side of the border had left him wondering about his identity. He described how, when playing on a Canadian hockey team against an American team, he felt somewhat "hypocritical" celebrating victories. When he was ready for a job, he said, he would decide which citizenship to keep and without wanting to "sound rude" or be interpreted as suggesting that Canada was a "bad country," he thought that he would probably "rather be an American" – a comment suggesting that he did not envision the possibility of positioning himself within a binationalized hybridity.

Not surprisingly, Dylan emphasized his particular enjoyment of the then binational Friendship Festival, because he felt that at this event,

Americans and Canadians could be seen "hand in hand, putting away citizenship and just wanting to have fun." For him, the mood of the festival embodied a positive bi- or even post-national "togetherness," that contrasted with the rest of the year when, he claimed, people on the Canadian side would make negative comments about "those damn Americans." At the festival, he claimed, "none of that matters because we're there because we're friends," something of great personal importance to him.

Dylan's struggles supported the speculation of Jim (who was pursuing dual citizenship) about a U.S.-based friend with dual citizenship, who he felt might have "ethnic identity problems," because while being "very proud" of being "half Canadian," when he was in Canada he was treated as an "American [while] ... when he's at home in the States, he's a Canadian."

While echoing the argument made by some others that Canadian border residents were less likely to stereotype Americans as for example, "Yankee hot dog eaters" or American-side inhabitants less apt to portray Canadians as always saying "Have some maple syrup, eh?" Jim also noted that border residents had to be particularly careful to avoid "saying the wrong thing in front of someone ... [that] you might not know ... is an American." He went on to personalize this with a story of how his own relationship to his American grandmother (married to a Canadian and living in Canada) became strained when he told her that American President Bush was the reason that he did not "believe in America," a comment, that "did not go over very well at all."

These accounts point to the ways in which Canadianness and Americanness could be experienced as simultaneously mixed, overlapping, in tension, and even mutually exclusive by Canadian border residents with close links to both sides of the river. While Ashley described herself to others as "American-Canadian, basically a mutt," these narratives suggest that for many, ethnicizing Americanness was difficult in the context of a Canadian nationalism marked by "the constant attempt to construct an authentic, differentiated, and bounded [Canadian] identity ... often ... through comparison with, and demonization of, the United States" (Mackey 1999, 145).[10]

At the local level, zero-sum constructions of national identity often articulated with constructions of inferiorized class and race in ways that further discouraged acknowledgment of binational families and/or American-Canadianness. This was clear in the comments of Sheila, who discussed why she did not publicly disclose her kinship links to the other side of the river. While denial of her American maternal grandmother made her "feel bad," she explained that having heard her American relatives described as "hillbillies," she just couldn't "bring herself" to admit her connections to this denigrated identity. Meanwhile, Rob described his sensitivity to what he experienced as anti-Americanism in the Canadian borderland, illustrating this by claiming that those on the Canadian side would say things like "Well, I might not have a job, I might not be doing very well, but at least I'm not like the blacks in Buffalo" – a comment that clearly revealed the classed and racialized constructions of border difference and inequality.

Such everyday experience and constructions stood in contrast to the ways in which cross-border marriages were invoked as emblematic of cross-border unity at two different Niagara-based 2003 rallies – one for and one against Canadian involvement in the U.S.-led invasion of Iraq. At each of these events, Canadian male speakers referred to their American wives. The pro-war speaker invoked his partner and their children and grandchildren as evidence that Canadian-American relationships could be "fruitful" ("Hundreds Show Support for American War Effort," *Niagara Falls Review*, 14 April 2003, A1), while the anti-war speaker referenced his American partner to underscore his argument that an anti-war position by Canadians should not be equated with anti-Americanism. Binational marriages in these political contexts seemed to be rhetorically invoked as emblems of unity across nationalized difference rather than Canadian-American hybridity.

Conclusion

Drawing on both border studies and studies of Canadian nationalism, this chapter used local press material and interviews to illuminate forms of everyday Canadian nationalism at the border. The discussion began by using press reporting to examine top-down elite discourses of both pre- and post-9/11 cross-border regionalism. These were then contrasted to more everyday accounts of how borderline Canadianness engaged with Americanness on the ground. While some Canadian border interviewees claimed that the "two sides" of the Niagara border were "pretty much the same," or that those on the Canadian side were Americanized, the more common constructions posited nationalized differences between Canadians and Americans even while borderline Canadians were simultaneously distinguished from co-nationals farther inland from the border. The strength of nationalized constructions of border space and identity was highlighted in the stories of American visitors and related expressions of anti-Americanism as well as the fraught positioning of dual Canadian/U.S. citizens, which illustrated the limits of a hybridized American-Canadian identity in the region. The salience of nationalized identity in the context of everyday binationalism was striking, as was the way in which nationalized Canadianness was classed and racialized. Constructions of Canadianness in terms of middle-class whiteness, however, invisibilized the diversity and inequality of Canadian Niagara and uncritically reproduced the exclusions and inequalities of white settler Canadian nationalism.

Chapter Six

Bordering Globalization at the Borderline

Introduction

As mentioned previously, my own thinking about how Niagara border life and bordering were globalized was sparked by reading about the tragic death of a child migrant at the Rainbow Bridge. The history of Canada/U.S. boundary making further highlighted the significance of globalized histories and political economies while more contemporary discourses of cross-border regionalism in Niagara often invoked globalization as a reason for greater cross-border cooperation.

In my formulation of this study of local engagement with top-down border projects, filtered bordering, and everyday nationalism, then, I wanted to ensure that I explored not only how border life was experienced and imagined in terms of the proximate U.S. but also how border residents in Niagara might understand their positioning within a wider, interrelated assemblage of borders in a globalized world.

In an effort to work against the pressures of methodological nationalism, and binationalism in particular, I highlighted the impact of globalization in the preamble to the interview schedule and made a point of asking interviewees directly about other borders beyond Niagara as well as global tourism and migration. A cluster of questions placed within a section titled "Globalization" also encouraged interviewees to reflect further on the issue and its impact on their lives as border residents (see the appendix). What is striking, however, is that the globalization discourse of the interview schedule was rarely taken up by interviewees. Discussion about other borders, global tourism, and migration in particular, often emerged only in response to more active prodding by the interviewer. As I discuss below, moreover, much of the more "global talk" was couched within a nationalist framing that remained predominantly focused on the relationship between Canadianness and Americanness. Keeping in mind these constraints, this chapter draws upon the more limited global talk of the interviews as well as local press materials to examine how border residents experienced and imagined themselves within a globalized world. I begin with interviewee accounts of globalized Canadianness and then turn to those accounts relating to global tourism in Canadian Niagara. I then focus in some depth on the phenomenon of global migration at Niagara, paying particular attention to local press reporting on unauthorized border crossings as well as pre- and post-9/11 refugee migration. Following further discussion of how national and global came together in interviewee accounts of local bordering, I conclude the chapter with a brief exploration of alternative counter-hegemonic border activisms and imaginings.

Globalized Canadianness

When directly asked about their experience of border crossings beyond Niagara, only a quarter of the interviewees indicated that they had travelled beyond the U.S., and few of these offered commentary on their global border crossings.[1] Brenda's reference to borders in Europe being "stricter" [than Niagara] because she "had to show [her] Canadian passport … [and] declare everything," and Amanda's story about what she considered to be the traumatic experience of having her luggage searched upon entry to Mexico were atypical.

While there was little commentary about other border crossings, a theme developed by those who had travelled beyond the U.S. as well as those who had not was that of globally welcomed Canadianness. As Ann put it, "People [globally] … assume Canadians are really nice people, they're liked around the world … If people know you are Canadian, then they are very friendly towards you," while Barb added that Canadians were globally "stereotyped" as "accepting and multicultural and … polite," while Americans were viewed as "more pompous … arrogant and not as accepting of things." Claims that nationalized character traits made Canadians internationally welcome were often, as in this case, linked to additional claims about putative global anti-Americanism. According to Bill, for example, "Everybody in the world thinks Canadians are just wonderful" because "Canadians always help people out [whereas] the Americans always step on people." The positive global reception of Canadianness versus Americanness was echoed by dual Canada/U.S. citizen June, who reported that during travel in Europe, she felt her "Canadian attachment ... was embraced a bit more."

Stories of global pro-Canadianism closely parallel Kymlicka's (2003) discussion of how the dominant Canadian nationalism in this period was marked by assertions that Canadians were globally recognized as "good citizens of the world" relative to Americans (359). In the interviews analysed here, such claims were accompanied by references to the importance of making one's Canadianness visible, given the putative tendency of global Others to mistake Canadians for U.S. citizens. Jessie reported, for example, that when she and her sister were in Europe, once "everyone ... recognized that we were Canadian ... they were so excited ... 'You're not Americans? Oh, wonderful.'" Dave mentioned that his vacationing parents would "always wear little Canadian pins on their shirts, because if people [thought] they [were] Americans they [got] treated differently ... not in a good way." Yvonne, who had worked as a missionary in the Caribbean, recounted, "When we said that we were Canadian, people treated us a lot differently [than they would Americans]," and Rachel, who did similar work in Mexico, reported dealing with presumptive Americanness by always clarifying, "'I'm Canadian' ... [because] there's an abrupt change in how I'm treated" (see Kymlicka 2003, 360, for further discussion).

Such descriptions of globally welcomed Canadianness by Canadian Niagara interviewees obscured not only the fact that global visa and migration regimes produce relatively frictionless global mobilities for (especially white) Canadians but also global critique of Canadian economic, political, military, and cultural imperialism abroad (e.g., Razack 2004). Assertions of a positively essentialized global Canadianness that contrasted with a negatively essentialized global Americanness reproduced and reinforced the kinds of nationalized border talk discussed in chapter 5.

Global Tourism

Another way in which globalization emerged in the interviews was in discussion about global tourism at Niagara. Some clearly felt that exposure to global tourists beyond visiting Americans was central to their identity as Niagarans. Mary, for example, described feeling "very special" growing up in a region that attracted so many nationalities to "a wonder of the world."

Some went further, linking the experience of hosting global tourists to the development of a degree of cosmopolitanism among Canadian border residents. Nancy was one of those who described how her own youthful employment in Niagara tourism provided "opportunities to have little cultural exchanges" that "opened up my eyes to the potential of the world." Tom, referencing summer employment that involved dealing with tour groups from Japan, Germany, France, Spain, and China, explained how "chatting" and "befriending" such visitors reduced stereotyping, because "it's easier to accept the stereotype if you've never met anyone of the other nationality or … culture." Anne felt that her "horizons" had been broadened through exposure "to different parts of the world," while Jason celebrated what he saw as positive interactions "between … people from other parts of the world … in Niagara." For Erin, contact with "different cultural groups" made Canadian Niagara "more multicultural" than "somewhere … like northern Canada," and Paula also described Niagara as "very multicultural," because "you see a lot of different ethnicities and stuff … tourist-wise."

While personal tolerance for the cultural diversity linked to global tourism was emphasized, there were also references to allegedly less tolerant views of fellow Canadian border residents. Ashley claimed, for example, that fellow employees in the tourist industry were "biased" when it came to "people of other cultures" (French-speaking visitors in particular), while Brenda, commenting on the increased numbers of Japanese visitors, noted that while she was raised "to embrace different cultures," other residents were "ignorant" and "bordering on racist," despite the fact that there were "a lot of different cultures coming in" to Canadian Niagara.

Expressions of tolerance and intolerance could be combined through differentiating among global visitors. While keen to "avoid stereotypes," for example, Ashley noted that she was most impressed by British visitors, because "even the young teenagers would call you 'dear' all the time" and were "very polite." Amanda also noted that the only enjoyable tourists were "English customers … from Britain or Australia." Both of these interviewees were more critical about other tourists. Ashley described wealthy Japanese spending "two, three hundred dollars at a time" on "tacky items," while Amanda complained about the behaviour of "Asians," most of whom "didn't speak English," and characterized "East Indians" as cheap, "messy," and "loud," with overly large families. The intersecting discourses of race and class were also apparent in Hannah's contrast between "Europeans" (who were "so friendly") and "Orientals" (who "didn't care what they were paying"). More favourable assessments of English-speaking, primarily white British, Australian, and broadly European visitors, I argue, was linked to a white-settler Canadianness that constructed these visitors as closer to an imagined core Canadianness. In contrast, other tourists were more clearly identified as "outsiders" who were then identified as the source of Niagara multiculturalism and/or Niagara racism. The construction of visitors as outsiders in these ways simultaneously constructed the local as monocultural and/or white thereby once again obscuring the historical and contemporary realities of a multicultural and multiracial Canada and Canadian Niagara.

Global Migration

Another way in which the interview schedule addressed globalization was through questions about global migration at Niagara. As it turned out, some interviewees also raised the topic of global migration in other contexts, including during broader discussions of Canadianness. Sophie, for example, suggested that while immigration was needed to address Canada's demographic challenges, she was concerned that – unlike in the past when her own European parents arrived and, according to her, "it was really difficult ... to come in ... [because] you had to have a certain net worth," – the current Canadian immigration system was allegedly "not quite fair," because it "let in people that don't have anything and then they use the resources that are here." The experience of living in a wealthy country like Canada, she felt, meant that "you want to protect yourself" from the "scary" scenario of inward migration from poorer countries.

Amanda also believed that immigration should be more restrictive. In making her points, she deflected anticipated critique, claiming, "I hate saying this but ... I have to be honest" before stating, "I hate the idea of people coming to Canada from other countries just because ... they can't get into the States ... it's so much easier to get into Canada and I hate that. I hate that I can't get a job because [the needs of] all these minorities have to be met."

In this constructed scenario, non-American migrants were entering through what was described as less restrictive (relative to the U.S.) Canadian border. Such migrants were then conflated with "minorities" within Canada whose putative claims (within the labour market in particular) allegedly trumped those of non-minoritized Canadians like herself. Amanda believed that multiculturalism was a "huge part of the Canadian identity," but that it was now at "the point that we don't even know what Canadian is anymore, because of all the people coming over and it being so easy to cross the border and to get in." In this account, a putatively open Canadian border was the source of poor and unproductive global migrants who, upon border crossing, transformed into "minorities," who, Amanda suggested, threatened both her economic security as well as a broader Canadianness.

Anna was familiar with the views expressed by Amanda because, as she explained, she had friends and family members who would complain about the lack of "our own Canadian culture" and suggest that some immigrants "should just get out ... and we'll get back to our Canadian roots." Even before 9/11, she reported, acquaintances would say for example, "A lot of these Pakis ... should ... just go home." She contrasted her own more positive views of the "sometimes confusing" diversity of Canadian culture, produced by the fact that "our roots are elsewhere," with the "ignorant" and "intolerant" positions of family and friends, attributing the latter to their being "too sheltered" and not having experienced global travel.[2]

Brenda, while claiming to be proud that in Canada "we try to accept people's cultures and their way of life and let them still live that way here," described herself as sufficiently shaken by 9/11 to be thinking instead about "how cultures are different" and, as a result, how "maybe it is better to have them [immigrants] melt into our way of life than to still keep their separate identity." Anne, however, felt that Canada/U.S. border harmonization should be resisted, as this would threaten a Canadianness marked by border openness to the world. According to her, "Canada has this welcoming appeal that we try to promote with our border," and unlike the U.S., she argued, "we give people a chance." As some of these examples indicate, discussions of global migration and the Canadian border were framed within a nationalized discussion of Canadianness versus Americanness.

Of the six interviewees who were immigrants to Canada (including two from the U.S.), the only personal account of migration came from Joe, who shared his dramatic "first border crossing," a perilous unauthorized crossing as a child refugee escaping his country of birth. Joe's unauthorized border crossing was not at Niagara, but it had parallels with the 1998 case that had first directed my attention to global migration at the local border. His account of this childhood crossing and subsequent familial migration to Canadian Niagara was an important reminder that global migration was part of growing up for some border residents. Beyond these personal experiences, interviewee discussions of global migration at Niagara focused on refugee arrivals and unauthorized border crossings. The sizable labour force of temporary migrant workers in Niagara (largely from Mexico, Central America, and the Caribbean) was not mentioned in this context (perhaps because the research was focused on the territorial border and these workers arrived by air).

During the research period, the Peace Bridge was a major refugee entry point to Canada, and some interviewees described childhood involvement in refugee-targeted charitable support and reception activities. Jason, who as a child had attended a private, religiously affiliated school on the U.S. side, commented on how his school had offered temporary education to children who were waiting for their "permit to be refugees or immigrants" and/or "to get into Canada." Jim was aware of refugees from "Third World countries," because he and his mother would help them get settled into accommodation. Dylan described raising money and offering other forms of support to refugees through his church, and Kerrie, who had volunteered with families who were "waiting to go to their trials to see if they can stay here [in Canada]," offered an unusually detailed and sympathetic account of the anxiety experienced by one family waiting for a decision on their refugee claim and expressed concern about the families who, fearful of an unsuccessful hearing would "disappear" to Toronto: "We never hear from them again."

Like others, Kerrie also learned about global migrants arriving in Niagara from an acquaintance who worked at the border. From this friend, Kerrie explained, she became aware of "families ... that illegally cross or have been living in the States illegally for years." Dale also described how border workers had told him about migrants who appeared without passports and were "obviously lying about how they got there because they [were] desperate to get in." He felt that many in Niagara were in fact unaware of what he considered the high number of refugee claims made by people arriving at the Canadian side of the Niagara bridges. Lisa explained how her own janitorial work at the border had given her a "negative view" of a system that allowed refugees to "come over ... in herds." While acknowledging that she didn't "know how the [refugee] system [worked]," she felt that it was "a joke" how many were trying to "get into ... Canada" and believed that many refugee claims were fraudulent, because she had "heard the lies ... through immigration ... behind-the-scenes stuff." She added that greater border control was needed, given the high volume of "refugees and immigrants coming through." Ed also expressed concern that "anybody that shows up at our border, if they're ... a refugee or running from political persecution, automatically they're given temporary status." This, he felt, was part of the larger problem: "We let anybody into our country that wants to come in."

Such negative attitudes towards global migrants at Niagara were referenced by Dale, who was critical of what he considered to be the "hypocrisy" of local residents who wanted "easy access" across the border for themselves – and if they encountered "any hassle whatsoever" would say, "What the hell is going on here? Let me through, let me through!" – but at the same time wanted to "tighten" the border "for immigrants and refugees," a position that he linked to an "underlying racism in Niagara." As these accounts reveal, both the broader Canadian and local Niagara border could be constructed as problematically permeable when it came to global migration.

Unauthorized Crossings Pre-9/11

In chapter 2, I described some of the historical connections between increasing Canada/U.S. border regulation, restrictive and racially exclusionary immigration regimes, and local involvement in smuggling migrants across the Niagara River. I also described previously how press reporting on the death of an infant migrant led to my own realization that the phenomenon of unauthorized migration studied at the Mexico/U.S. and other borders (e.g., Nevins 2007; Spener 2009; Khosravi 2010) was part of contemporary Niagara. Struck by the relative silence on this topic within the border-related policy and scholarly work on the Canada/U.S. border at the time, I wanted to learn more about Canadian border resident experiences with, and understandings of, dangerous unauthorized crossings that appeared to have the potential to unsettle hegemonic constructions of a benign Canada/U.S. border.

A number of interviewees shared knowledge of unauthorized crossings. Zak, for example, referred to the "elusive" entry to the U.S. by a local Canadian man whose criminal record precluded authorized entry, but most described not local but global unauthorized crossers. Dave had heard stories about global migrants "crawling across the top [of the bridge]," and Kevin was aware of would-be crossers "trying to hop trains." Lisa shared her knowledge that Canadian border workers had found "a couple of refugees smuggled in," and Avery referred to crossers "hanging on to the bottoms of the trucks." Kerrie's friend at the bridge had also told her "stories about ... people ... [who] would hide under transport trucks," and Hannah said that even during post-9/11 "high alerts," it was still a "frequent thing" for truck drivers to "smuggle over immigrants." Sophie had seen a TV show about people from Costa Rica and Guatemala taking "risks" trying to reach Canadian Niagara and knew that some unauthorized migrants had tried to cross at the Whirlpool Rapids Bridge. She had also learned from the local press about "Chinese immigrants ... strapped underneath ... a truck ... [who] suffocated in the back of a semi."

As these references indicate, many of these interviewees were aware not only that unauthorized migrant crossings occurred but also that these could involve migrant injuries and fatalities. Jason, who came to know of these realities because his father was an emergency worker, recounted that one time "two people from Honduras [tried] to cross over and one of them died in the crossing," while "another time there was somebody from Guatemala that tried to cross in and this again was unsuccessful." Several referenced the particular dangers of river crossings. Debbie recounted hearing about "some man, I think he was Chinese. He was trying to come to Canada from the States and I remember hearing about ... how he...fell back into the river and that really scared me." Ed, who in the course of his employment became aware of the retrieval of human remains from the Niagara River, also recalled that

> Chinese immigrants ... trying to get from Canada to the United States ... bought an inflatable pool ... [and] tried to paddle across the lower Niagara River ... [but] there were ... hunks of ice in the lower river at that time and the raft ... sank.... One of them drowned.

Chuck cited the U.S. media as the source of his awareness of mostly "minorities and ethnicities," who risked their lives swimming across the river. He concluded from such reports that it was easier for "ethnic minorities" to enter Canada than the U.S. and felt that this indicated that Canadians were "more welcoming to other people." He added, however, that it was his view that the degree of Canadian welcome should be determined by would-be migrants' "purpose in the country," and that this should be withdrawn "if they're just trying to sneak in." Others shared a concern about links between unauthorized crossings and criminality. Jessie, for example, felt that one of the drawbacks of border life was greater criminality linked to drug and people smuggling. As she recounted, "You hear all these stories of people smuggling things over, people ... sneaking in or hiding in transport trucks and going over the border ... that's definitely a predominant aspect of a border community."

Sophie suggested that locals were implicated in this smuggling-related criminality when she noted that the Niagara region had a reputation for both police corruption and "organized crime," including the smuggling of "alcohol or drugs or ... people," and Mark, who had heard that "immigrants" denied legal entry would "sneak over in the back of trucks ... and then try to work over here," decried such activity along with "smuggling drugs or clothing, stuff that was maybe stolen." The proscription of human and other forms of organized smuggling as "criminal" in these discussions clearly contrasted with the apparent widespread tolerance, described in earlier chapters, of border resident involvement in small-scale smuggling for household social reproduction.

As indicated, many interviewees had learned about unauthorized crossings from the press, and my review of the Canadian *Niagara Falls Review* revealed recurrent reporting on the phenomenon. Some of the reports related to crossings by local Canadians and U.S. citizens. These crossings, however, were rarely described with the "illegal" terminology applied to other nationalities. Coverage of the interception of a Canadian man attempting to enter the U.S. in a rowboat and of a local Canadian child and teenager who walked undetected across bridges in separate incidents, for example, avoided such a framing.[3] That unauthorized crossings by local Canadians and U.S. nationals could even evince sympathetic local press coverage was apparent in reporting under the headline of "True Love Knows No Borders: American Man Now Deported Eight Times for Visiting His Wife" (*Niagara Falls Review*, 17 July 2003, A5), on the case of a man from Niagara Falls, New York, deported after visiting his Canadian wife, who could not join him in the U.S. due to her criminal conviction. Other press stories also humanized unauthorized Canadian and American crossers by focusing on how their romantic or familial reunification goals were being thwarted by border enforcement (e.g., "St. Catharines Bride Fighting U.S. Immigration Rules," *Niagara Falls Review*, 20 August 2003, A2).

The absence of illegalizing terminology applied to the unauthorized crossings and everyday smuggling by locals meant that press references to "illegal" crossings and "smuggling" were most consistently linked to non-Canadian or U.S. nationals, often constructed as "illegal aliens"– a demonstration of how "illegality" and "alienality" are "socially, culturally, and politically constructed" (Chavez 2007, 192).

This language also reflected the fact that local press reporting on the latter category of "illegals" relied almost exclusively on U.S. border enforcement for (often verbatim) accounts of U.S.-bound migration. Such reports, while frequently highlighting the danger and deaths involved in such crossings, distanced Canadian Niagarans from the issue, because they involved U.S.-bound crossings. Keeping in mind the ways in which local Canadian border press framing of the issue reflected U.S. border enforcement perspectives, I nonetheless draw upon these press reports to convey something of the phenomenon itself, while also considering how it was portrayed to Canadian border residents.

References to unauthorized crossings by presumptively global migrants in the *Niagara Falls Review* data set began with February 1999 reporting on a dramatic incident involving a U.S. customs agent's discovery of four Chinese women under a U.S.-bound truck on the Queenston-Lewiston Bridge ("Desperate Border Run Stopped at Queenston-Lewiston," *Niagara Falls Review*, 1 February 1999, A1), a story that was picked up by the Canadian and American national press.[4] Following this incident, the issue of "illegal migrants" at Niagara reappeared in local and national reports when six U.S.-bound individuals, also identified as Chinese, were intercepted while attempting to crawl across the Whirlpool Bridge just over a week later.[5] The local reporting on this second incident referred to a history of migrants drowning in the Niagara River.

Two weeks later, the Canadian national paper, the *Globe and Mail*, reported on the death of a young woman described as originally from Peru, who had died of injuries sustained while jumping off a moving railway car after being spotted by U.S. Border Patrol at Niagara. This young woman had been part of a group of six that included her brother (a Canadian citizen) and two Mexicans. The woman and her brother reportedly had relatives in the U.S., while one of the Mexicans was attempting a re-entry ("Woman Killed in Bid to Sneak into US on Train," *Globe and Mail*, 26 February 1999, A3).

The *Niagara Falls Review* story of early February had quoted U.S. immigration officials describing Niagara as a "smuggler's paradise," and the *Globe and Mail* report of late February suggested that the region was experiencing an unprecedented number of unauthorized crossing attempts, as indicated by interceptions of people from China, Colombia, El Salvador, Barbados, Mexico, Peru, and Africa. The Niagara flows were attributed to greater border enforcement at the Mexico/U.S. border and at the much closer Canada/U.S. crossing at Cornwall, Ontario/Massena, New York, where, it was reported, thirty-five "illegal Chinese aliens" had been arrested two months earlier and allegedly "at least 3,600 others [had been smuggled across] in the previous 2 years." More effective enforcement at these sites, it was suggested, was redirecting irregular migrants to Niagara ("Desperate Crossing," *Globe and Mail*, 26 February 1999, A3).

By the following summer, the *Niagara Falls Review* quoted a spokesperson for the Buffalo Border Patrol arguing for video surveillance cameras along a section of the Niagara River to combat the smuggling of "contraband," which the news story clarified included "alcohol, cigarettes and illegal aliens." The Buffalo Border Patrol spokesperson reportedly claimed that while there had been only about twelve arrests in 1999, as many as "300 people probably made the crossing, some using very simple rubber rafts" ("Lewiston Residents Reject 'Big Brother,'" *Niagara Falls Review*, 15 August 2000, A3). In a *Niagara Falls Review* story of February 2001, it was reported that since October 2000, the local U.S. Border Patrol had intercepted forty-four U.S.-bound people from twenty-two countries who were attempting to walk over Niagara railway bridges ("Media Campaign Takes Aim at Human Smuggling," *Niagara Falls Review*, 9 February 2001, A1).

As indicated above, at least some of these cases involved attempted re-entries by legal residents or citizens of the U.S. One attempted re-entry early in 2001 involved a male identified as originally from Zimbabwe. This individual had been living and working legally in the U.S., but after crossing the border to visit friends in Windsor, Ontario, he was barred from re-entry because he reportedly tried to "sneak a Zimbabwean friend into the U.S." With his visa revoked, this man made his way to the Niagara border region a few hours away and climbed underneath a U.S.-bound bus. U.S. officials at the Peace Bridge discovered his mangled body shortly afterward ("Stowaway's Body Found Wrapped around Bus Axle," cbc.ca, 3 January 2001).[6]

A *Niagara Falls Review* article that made reference to the above case included a U.S. official's claim that this was one of seven known deaths of people attempting unauthorized U.S. entries at Niagara in "recent years" ("U.S. Beefs Up Border," *Niagara Falls Review*, 19 January 2001,

A1), and fatalities were also referenced in the report on a local U.S. border enforcement campaign against "human smuggling" that included public service announcements featuring footage of "raging currents in the river, the dizzying heights of the Niagara Falls bridges and the unstoppable power of moving freight trains followed by the image of a body laying [sic] in the snow covered by a blanket and then being loaded into an ambulance" ("Media Campaign Takes Aim at Human Smuggling," *Niagara Falls Review*, 9 February 2001, A1).

As these news stories reveal, state preoccupation with "illegal" migrants at the Canada/U.S. border preceded 9/11, and – although on a much lesser scale than either the Mexico/U.S. border or European frontiers – the pre-9/11 Niagara border was a site of violence where, as at many other borders, migrants were experiencing "bodily what the 'securitization' of migration by the state means in practice; that is more risk, more insecurity and more unnecessary deaths" (Munck 2008, 1242). Despite considerable local reporting and suggestions in the interviews that many Canadian border residents did have knowledge of this more perilous aspect of the border, it remained a minor theme of the interview narratives. Such local silences supported more official silences and thus left uncontested the hegemonic images of a benign rather than a deadly Canada/U.S. border.

Refugee and Unauthorized Crossings Post-9/11

Some discussion of global migration at Niagara had emerged in interviewee references to their involvement in various forms of refugee support. Despite these involvements, none talked about the dramatic impact of post-9/11 securitization on local refugee flows. What was clear in the local press, however, was that following 9/11 there was an effective stranding of Canadian-bound refugees as a result of the immediate closure of the local Canadian immigration office – a closure that remained in place even while more general cross-border flows were being re-established ("Refugees Backed Up at Border," *Niagara Falls Review*, 26 September 2001, A3). Post-9/11 migratory flows were also affected by new restrictions on global mobilities and migration both at and beyond the border itself. The Canadian Immigrant and Refugee Protection Act that came into effect in 2002, for example, increased bilateral cooperation on interdiction, expanded powers of detention and deportation, and imposed new penalties on human smuggling. This was combined with the Canada-U.S. Safe Third

Country Agreement (signed in 2002 and discussed further below) to produce shrinking "spaces of asylum" and other forms of authorized entry to Canada, paralleling similar developments in other wealthy states (Mountz in Johnson et al. 2011, 66). At the Niagara border itself, unauthorized crossings were targeted by increased investment in border patrol agents, helicopter and marine surveillance, motion sensor cameras, scanning, and other technologies.

While press reports focused on so-called illegals in the post-9/11 period, they also made it clear that increased local securitization did not stop unauthorized crossings. In April 2002, for example, a *Niagara Falls Review* report on the apprehension of four U.S.-bound people identified as Chinese and Malaysian claimed that "after months of relative quiet, the Niagara River is again bustling with clandestine traffic as illegal aliens make a desperate drive to slip into the United States from Canada." A spokesperson from the U.S. Buffalo Border Patrol was quoted as saying that smuggling attempts were returning "to a more normal level" and that there might be a "backlog of people who want to come across." A newly visible bilateral cooperation was indicated by an unusual quote from a Royal Canadian Mounted Police spokesperson, who stated, "We've had several incidents involving rafts and Asians on the river" ("Smuggling across River on the Rise," *Niagara Falls Review*, 25 April 2002, A2). A few days later, seven U.S.-bound men identified as Costa Rican (one of whom was identified as a previous resident of the U.S., with children there) were intercepted on a railway bridge ("Seven Arrested after Attempt to Enter US," *Niagara Falls Review*, 30 April 2002, A3).

In December 2002, a *Niagara Falls Review* report about a couple from the Dominican Republic who managed to swim back to the Canadian shore after their dinghy capsized, quoted a U.S. Border Patrol spokesperson as saying that, despite the cold weather, such attempted crossings were happening "all the time ... [and] as long as they [crossers] can get out to that water, they're going to keep doing it" ("Pair Survive Spill in Niagara River: Couple Tried to Cross Lower River in Dinghy," *Niagara Falls Review*, 4 December 2002, A1). A U.S. migrant advocacy report meanwhile cited an August 2002 *Boston Globe* story, which stated that U.S. authorities had found nineteen bodies in the Niagara River in the eleven months following 9/11 and that most were "believed to be undocumented immigrants" (Liebmann 2002, 5).

Dangerous crossings, then, continued and may have increased after the initial post-9/11 hiatus. From January to May 2003, U.S. Border

Patrol agents reported intercepting eighteen "illegal immigrants" using "rafts or other flotation devices." One of the male crossers was described as "sneaking across the river ... on an inner tube with duct tape wings" before being hauled out of the river by two fisherman ("Residents Recruited for Border Enforcement," *Niagara Falls Review*, 31 May 2003, A5). In 2004, nine U.S.-bound "illegals" (although one was a U.S. citizen) – were arrested after trying to cross on a raft.[7]

At the end of 2004, the issue of unauthorized crossings became explicitly linked to Canada-bound refugee crossings in the context of reporting on the local impact of the bilateral Canada-U.S. Safe Third Country Agreement, which required those travelling through the U.S. en route to making refugee claims in Canada to instead make their claim in the U.S. The negative impact of the agreement had been highlighted by refugee advocates, who protested in Niagara and elsewhere, arguing that the U.S. was not, in fact, a "safe country" for many from Central and South America, whose claims for asylum were more likely to be rejected there ("Refugees' Rights Protest Held at Border," *Niagara Falls Review*, 10 May 2003, A3). A court challenge by national non-governmental organizations, notably the Canadian Council for Refugees, Canadian Council of Churches, and Amnesty International, however, was ultimately unsuccessful.

The negative impact of the Safe Third Country Agreement on authorized migration for refugees at Niagara was clear when, in December 2004, the month before implementation of the agreement, the Canadian side of the Peace Bridge saw the unprecedented arrival of 1,800 asylum seekers – leading the Peace Bridge Newcomer Centre on the Canadian side to send out an urgent call for shelter, food, and settlement services to the local Canadian Red Cross, Salvation Army, Fort Erie Multicultural Centre, Welland Heritage Council and Multicultural Centre, and Niagara Regional government.

Local press reporting on the impact of the agreement included predictions by refugee supporters that the result would be unauthorized crossing attempts by refugees now denied authorized entry ("Refugee Proponents Predict People Smuggling," *Niagara Falls Review*, 30 December 2004, A1). One Buffalo-based refugee advocate, for example, was quoted as saying that migrants would try to swim or to scale "the underside of any of the bridges across the Niagara River" ("Vigilance Needed at Border," *Niagara Falls Review*, 4 January 2005, A1). A report published fifteen months after the implementation of the agreement noted that there had indeed been an increase in "extremely dangerous" unauthorized crossings to Canada (Harvard Law School 2006, 3), and an internal Canadian Border Services Agency report later acknowledged that across the Canada/U.S. border as a whole, the Safe Third Country Agreement "inadvertently increased numbers of unauthorized entries into Canada between ports of entry by migrants who wish to avoid being turned back at the border" (Arbel 2013, 72).[8] The impact of greater securitization, combined with the Safe Third Country Agreement, was also linked to an apparent reversal in the direction of unauthorized crossings, as the limited official figures relating to interceptions between ports of entry across the Canada/U.S. border pointed to more Canadian- than U.S.-bound apprehensions by 2007 (Canada–United States Integrated Border Enforcement Team Threat Assessment 2010).

Local reporting, however, continued to uncritically report U.S. official claims that border security was reducing unauthorized crossings rather than exacerbating them. In one story, a Buffalo Border Patrol spokesperson claimed, for example, that the introduction of new scanning technology for trucks and rail containers had reduced the "smuggling of illegal aliens" from over a dozen attempts involving "50 aliens being captured" in 2004, to just five arrests for "illegal human trafficking" in 2005. This statement, which characteristically collapsed the legally distinct activities of human "smuggling" and "trafficking" (see Ford and Lyons 2012) as well as those being "smuggled" and those doing the "trafficking," was ambiguous, but the headline "High-Tech Borders Reduce Alien Smuggling" was not (*Niagara Falls Review*, 6 December 2005, A1). The empirical basis for the headline's claim, however, was unclear, as official figures (e.g., for apprehensions between Niagara ports of entry) were not publicly available. Meanwhile, unauthorized crossings clearly continued, and the dangers of making such attempts without smuggler assistance (which security agencies termed "unfacilitated" crossings) were highlighted by the drowning death of one of two men from the Czech Republic who attempted to swim across the Niagara River ("Man, 34, Presumed Drowned in River Crossing," *Niagara Falls Review*, 13 August 2005, A3).

While an emphasis on "spectacular" border tragedies risks diverting "attention from the 'brutal mundanity' of borders" (Anderson, Sharma, and Wright 2009, 14), interviewee and press reports revealing dangerous and sometimes fatal unauthorized crossings at Niagara demonstrate that the Canada/U.S. border at Niagara, like many other territorial border sites, is "a particularly acute site for the violation of human rights in the era of globalization" (Sutcliffe, as quoted by Munck 2008, 1239). How such "localized, seemingly aberrant phenomena" were linked to "broader enforcement trends happening across international borders" (Mountz 2010, 169), however, went largely unacknowledged, and constructions of a benign Canadian border were reinforced by a focus on U.S.-bound flows (despite the important discussion of Canada-bound refugees). This downplayed Canadian complicity in an issue identified as primarily American.

A further local distancing from the violence of the border was evident in a tendency to blame the victim, as unauthorized crossers were portrayed as having put themselves at physical risk, given the allegedly self-evident dangers of their efforts. In press reports, for example, "illegals" were described as "risking death" or "taking their life in their hands" in "death-defying" crossing attempts. Crossers could also be characterized as "stupid" – as when Rachel mentioned that every few months she would hear on the news that someone had tried to swim across the river but added, "It wasn't like a Mexico-to-United States thing, like they were trying to start a new life … I don't know if people were trying to commit suicide or were just being stupid." Likewise, Tim described an unsuccessful attempt to smuggle "Asians" across the river in "a twelve-foot aluminum boat" as "bone-headed," noting that the smuggler "got stopped by customs and … thrown in jail. He didn't make it very far."

As the latter commentary suggests, there was a blurring in local discourses between unsuccessful migrant crossings and suicides or accidents. Some indicated in addition that, while there was local knowledge of river deaths at Niagara, there was a more systematic attempt by local leaders to avoid publicizing the phenomenon more widely. Dave, who had friends whose work included retrieving human bodies from the river, referred to a media "code of conduct," which entailed the non-reporting of "people who jump" (purported suicides rather than unsuccessful border crossers), adding that this "dirty little secret" was linked to the fact that "Niagara Falls has the largest collection of unidentified body parts in the world."

Sheila also shared her awareness that "a lot of people try to get over … they'll swim across and stuff." She then added, "At the end of the [summer] season they do the scooping at the end of the gorge, and they just pull all the dead bodies out and it's not something you see in the paper … people that don't live around here don't know about it."

That the phenomenon of "jumpers" could be linked to border enforcement was obscured in local media reporting. In June 2000, for example, a man travelling on a bus was turned back at the U.S. side of the Peace Bridge for failing to produce proper documents. The individual reportedly then took a cab back over the bridge towards Canada, but just before border marker, he exited the cab, climbed over the bridge railing, and dropped into the river below. Despite an ensuing boat and helicopter search, he was not located and was presumed drowned. In this case, where denial of border entry clearly preceded the "jump," a Canadian Niagara Regional Police spokesperson was quoted as asking rhetorically, "You've got to wonder what's going through a man's mind up there?" thereby downplaying the role of border enforcement in favour of speculation about the unfathomable mystery of individual psychology ("Search Fails to Find Peace Bridge Jumper," *Niagara Falls Review*, 6 June 2000, A1). Likewise, some years later, a Canadian Niagara Parks Police spokesperson, quoted in a local report of the discovery of two bodies within an hour of each other, noted that recovering bodies from the river was not "an unusual occurrence" but made no reference to the possibility of such discoveries being linked to attempted border crossings by unauthorized migrants ("Police Recover Two Bodies from River," *Niagara Falls Review*, 3 July 2004, A3).

In 2006, when the Ontario Provincial Police requested public help in identifying "unclaimed bodies," the website linked the large number of unidentified remains found in Canadian Niagara to the area's dangerous natural geography rather than an exclusionary border that led to deadly crossing attempts ("Website's Unclaimed Dead Found in Niagara: Many Bodies, on OPP Site, Found in City," *Niagara Falls Review*, 20 May 2006, A1).

Slack and Whiteford (2011) point out with regards to the Mexico/U.S. border that the U.S. Border Patrol highlights the "vulnerability of migrants to smugglers" rather than the role of states in producing migrant insecuritization (15). Similarly, in the U.S. border enforcement discourse reproduced in the local Canadian press and consumed by local residents, the violence experienced by unauthorized migrants at Niagara was generally linked to allegedly unscrupulous "smugglers" rather than differentiated bordering.

The report on the 1999 incident of the four Chinese found under the truck, for example, quoted the spokesperson for U.S. Immigration and Naturalization Service in Buffalo as claiming that the case revealed "what a callous disregard smugglers have for people" ("Desperate

Border Run Stopped at Queenston-Lewiston," *Niagara Falls Review*, 1 February 1999, A1), while another spokesperson was reported to have stated that "smugglers are not altruistic human beings ... they prey on human misery. All they care about is money, not their human cargo" ("Media Campaign Takes Aim at Human Smuggling," *Niagara Falls Review*, 9 February 2001, A1). A Canadian columnist likewise described "smugglers," not the crossers, as "the real criminals" ("Smugglers Real Criminals," *Niagara Falls Review*, 12 February 2001, A4). Repeated U.S. official references to "snakeheads," moreover, linked smuggling to non-national Others and organized crime, despite evidence of local border resident involvement (e.g., "Canadian Gets Probation for Smuggling Aliens From Falls," *Niagara Falls Review*, 10 November 2005, A1).

As Spener (2009) notes with regard to the U.S./Mexico border, crossing cases that receive media attention tend to be the failures involving "a death, an accident, or an arrest," and the accompanying negative media portrayals of those described as human smugglers or traffickers reflect border enforcement perspectives, as crossers and their assistants do not speak to news outlets (203–4). Border enforcement officials, meanwhile, may exaggerate links between small-scale smuggling and "organized crime, drug trafficking and terrorism" to ensure support for "ever-larger police and military budgets and infringement of basic civil liberties" (Spener 2009, 124–5). Far less visible is the complexity of a migration industry that offers migrants escape from, and movement through, violent geographies to hoped-for better lives, and in which some human "smugglers" may be family and/or friends (see Khosravi 2010; Van Liempt and Sersli 2013).

In Canadian Niagara, local press attribution of the dangers faced by crossers to the physical geography, their own psychology, or victimization by non-local "smugglers" and uncritically reproduced (especially U.S.) border enforcement self-portrayals of their interceptions of migrants' attempted border crossings as benevolent or chivalrous "rescues" (see Pratt and Thompson 2008). In the 1999 truck incident mentioned above, for example, the U.S. spokesperson voiced the opinion that the customs agent who discovered the four Chinese migrants "might have saved the women's lives" ("Desperate Border Run Stopped at Queenston-Lewiston," *Niagara Falls Review*, 1 February 1999, A1) and reporting on the interception of a man found treading water near the U.S. shore of the Niagara River quoted U.S. Border Patrol as claiming, "It was very fortunate for him that we were there because he could have very easily drowned" ("Man Plucked from Niagara River: Suspect Believed Trying to Enter US Illegally," *Niagara Falls Review*, 26 September 2002, A1).

Such claims were made despite indications that at least some of the "rescued" had engaged in their most risky behaviours (e.g., jumping off trains, entering the water, hiding on railway bridges) as last-ditch efforts to evade the intercepting "rescues" by border enforcement that would terminate their crossing attempts. Just days after the Safe Third Country Agreement had restricted authorized crossings, for example, the apprehension of two Chinese men who had tried to enter Canada by climbing a railway bridge on which they had "spent several chilly hours high above the Niagara gorge" was described as a joint Canada/U.S. "rescue" ("Two Nabbed at Bridge Trying to Enter Canada," *Niagara Falls Review*, 3 January 2005, A5). Descriptions of state-initiated terminations of migratory journeys as "life-saving" construct migrants as "objects of control, rescue, and redemption," thereby denying their autonomous "subjectivities, engagements, and actions" as desirous crossers (Anderson, Sharma, and Wright 2009, 8). They also serve to obscure the role of states in insecuritizing migrants through economic and political global displacement combined with restrictive migration regimes and, as in the cases of interception described above, instigating the actually "life-threatening" detention and deportation of migrants via "transnational corridors of expulsion" (Nyers 2010, 414).

National/Global Bordering

Of interest to this study is how local knowledge of unauthorized crossings was linked by locals to filtered bordering and constructions of everyday Canadianness. As Ford and Lyons (2012) note more generally, anti-migrant securitization has produced physically altered borderscapes of enforcement as well as altering "the ways that some borderlanders understand the nature of cross-border mobility" (451). In Niagara, the portrayal of greater surveillance and interception of unauthorized migrants as a benevolent project was important to official attempts to involve border residents in unpaid border enforcement. Post-9/11, U.S. border officials praised U.S. border residents for reporting "suspicious" river activity and asked Canadian border residents to be more active in this regard ("Border Crackdown Slows Illegal Entries to US: Citizens Playing Role After Sept. 11 Attacks," *Niagara Falls Review*, 10 November 2001, A1).

In 2002, the U.S. Border Patrol again urged Canadian border residents to provide volunteer river surveillance, suggesting that this would offer a benevolent form of protection for child migrants endangered in the course of accompanying unauthorized adult crossers ("Residents Thwart Alien Smuggling: Authorities Worry About Chances Being Taken with Children," *Niagara Falls Review*, 13 August 2002, A1). By 2003, the newly formed Canada–United States Integrated Border Enforcement Team more formally asked those living on both sides of the river (described as "500 pairs of eyes") to use a 1-800 number to report "suspicious activity" on the water ("Residents Recruited For Border Enforcement," *Niagara Falls Review*, 31 May 2003, A5).

Generally, interviewees supported U.S. border securitization that subjected those that they viewed as potentially dangerous global Others to greater scrutiny. As Ed (who had tried unsuccessfully to migrate to the U.S.) himself stated, the "Americans probably need it [a secure border] more than we do ... [because] everyone wants to get into the States." There were concerns, however, that more indiscriminate U.S. securitization problematically and unnecessarily restricted Canadians along with the allegedly more appropriately targeted global Others (e.g., the "Cuban ... or Mexican people" identified by Sophie). For Bill, the possibility of the U.S. putting "fences up along the [Canadian] border like in Mexico" was a "scary thought," and a 2005 *Niagara Falls Review* editorial about U.S. plans for increased fencing at the Mexico/U.S. border asserted proactively that "Canadians don't need – or want – to be fenced in." This editorial noted that while "[Canadian] Niagarans [were] well aware [sic] illegal aliens have been caught crossing the shared border with New York State" such cases were "tiny compared to the issue faced in the U.S. at the southern border crossings" ("Editorial: Don't Fence Us In!" *Niagara Falls Review*, 20 December 2005, A4). Here, Canadians were positioned outside the collectivity of "illegal aliens," a category reproduced as an appropriate target of increased U.S. border securitization but not enough of an issue to warrant a fence.

Zak's concern that the U.S.-led post-9/11 securitization – which had produced "barriers up everywhere" – was only going to increase led to his suggestion that a proactive elimination of the local border by Canada might be the way to forestall the negative impact on Canadians. His interest in dismantling the Niagara border in favour of a "Fortress North America," however, was not voiced by any other interviewees. Even when I directly raised the "perimeter security" concept of a common North American border floated by U.S. President Bush in 2001 but rejected by then Canadian Prime Minister Chretien, this garnered no additional support from interviewees – a reaction that pointed again to the strong investment by border residents in the local border as a marker of nationalized difference and autonomy.

Brenda, who after 9/11 experienced a new sense of affinity with Americanness vis-à-vis imagined dangerous "other cultures" that she described as changing her sense of "us" (Canadians) versus "them" (Americans) to a joint sense of "us" (Canadians and Americans) versus the world, nonetheless expressed opposition to the proposal of "one big North America." Like others, Amanda also expressed concern that "if there was just one North America … we'd [Canadians] … just get lost in it." Despite having grown up with free trade and, in the case of some, a new post-9/11 sense of commonality with the U.S., interviewees showed little evidence of engagement with concepts of North Americanness.

For some, then, 9/11 led to a heightened interest in independent Canadian bordering against putatively threatening global flows. As dual citizen Ashley explained:

> I've always loved the idea of being somewhat of a free country. You can come and go easily but I guess considering that the world's becoming smaller because of globalization, you would need to have some tight security to make sure that something like this [9/11] doesn't happen again.

Avery also felt that in the post-9/11 context the reliability of the U.S. as a bulwark against a dangerously globalized world was reduced, because while Americans were "supposed to kind of take care of Canada … they've become so vulnerable … they're really shook up … they can't really even deal with their own problems. How are they going to deal with ours too?" References to global terrorism, then, were linked to reflections on the binational relationship, with some, such as Emma, feeling that while economic ties to the U.S. were a "good thing," to be too closely associated might increase Canadian geopolitical insecurity.

The complexities of Canadian border residents' relationship to local bordering can be further illustrated by the case of Amanda, who saw post-9/11 "heightened awareness and protection of the borders" as "a good thing." Her support for greater bordering against global Others came less from fears of terrorism, however, than from her identification of global migration as a threat to her own economic aspirations and an imagined Canadianness. These views, however, coexisted with her own extensive local border crossings for shopping and family reunions with nearby U.S. relatives. Her boyfriend was employed as a cross-border truck driver, and she herself had worked at a U.S. summer camp across the river, where the staff came "from all over the world." Some of her friends among this global staff had crossed the local border to visit her in Canada, and, like these friends, Amanda had a personal history of global leisure travel and hoped to enjoy more of this in the future. Amanda's support for re-bordering vis-à-vis global Others – imagined as posing an economic threat to her future employment prospects – therefore coexisted with considerable personal investment in local and global cross-border mobilities. Significantly, these included the local border crossings of Canadian friends who were paying higher tuition fees for teacher training on the U.S. side of the Niagara River, where entry requirements were lower but teaching credentials recognized on the Canadian side could still be obtained. While Amanda expressed frustration that her U.S.-trained friends would be eligible to compete with her as a Canadian-trained teacher for limited jobs in Ontario, the cross-border mobility of these friends, despite posing a direct challenge to her own career aspirations, remained less problematized than those of imagined global Others.

Alternative Borderings?

As I have demonstrated, top-down border projects – both central-state and more regionally based – were engaged by Canadian border residents with varied degrees of acquiescence, support, critique, and transgression. Before ending this chapter, I want to reflect more deeply on the ways in which borders, while often "sites of intense state violence and inequality," may also be "generative sites" where border residents "experiment with new cultural, political, and economic forms" (Galemba 2013, 282). The idea of the border as a "generative site" led me to attend to the ways in which the border sometimes became a stage for various forms of civic activity, including what might be recognized as forms of counter-hegemonic politics offering degrees of challenge to the weak democratic political processes governing North American border projects (Gilbert 2007, 82).

Niagara Falls Review reporting revealed that, while the Canadian Niagara border site was often used by provincial and national politicians as a backdrop for such events as the one-year commemoration of 9/11 and policy announcements related to securitization and immigration ("Three-Minute Silence at Bridge Will Mark Sept. 11 Tragedy," *Niagara Falls Review*, 7 September 2002, A3; "Border Security Tightened: Caplan Unveils 'Maple Leaf' Resident Cards for Immigrants," *Niagara Falls Review*, 13 October 2001, A1), it also served as a venue for a variety of activities aimed at contesting forms of elite political and/or economic power. These examples of what might be understood as alternative border-sited activities and activisms illustrated how borders can be sites of both security and dissent (Amoore in Johnson et al. 2011, 64).

Along with the decades-long direct challenge to settler bordering by Indigenous protestors and their allies previously mentioned, for example, there was *Niagara Falls Review* reporting on a range of other border-sited protests, including labour-related actions by workers (i.e., federal public sector employees, Bell employees, and customs officers), coalition demonstrations against free trade and capitalist globalization, anti-war protests against the invasion of Iraq, and refugee advocate demonstrations against the Safe Third Country Agreement.[9] These actions sometimes used the threat of disrupting cross-border flows as a means of amplifying their message. Dale's account of being told by authorities during his own participation in one of these forms of border-sited activism, "Don't go on the bridge or you will be arrested," however, was a stark reminder of the limited space available for challenges to state power at the border.

Along with noting examples of the border site being transformed into a venue for counter-hegemonic protest, I also listened closely in the interviews for evidence of alternative border-related visions. I was interested, for example, that a few of the interviewees engaged with the idea of globalization in ways that appeared to challenge more dominant nationalist and/or anti-global Other frameworks. Tom, a self-described socialist, for example, critiqued "nationalism" for producing "stereotypes" and "segregation" and capitalist "globalization" for what he saw as its negative economic impact on his Niagara community. His preferred political vision was of "one world ... under one umbrella." Mary, who critically linked globalization to "multinational capital," also critiqued nationalism for obscuring of the ways in which both U.S. and Canadian citizens were "oppressed" by "a super ... billionaire group of people." Dale, meanwhile, explained that involvement in union activism and subsequent political education had led him to reject an earlier tendency to say "I hate Americans" and to adopt a "working-class" rather than nationalized identity. Joe, coming from a religious perspective, also described how he rejected the anti-Americanism of some of his friends in favour of a one-world theology that saw borders as "artificial." Such politically and religiously framed perspectives offered alternatives to both the dominant and the more common everyday border resident discourses of nationalism and globalization.

Attention to counter-hegemonic border-related practices and imaginings highlights further the importance of a critical Border Studies that documents such alternative everyday border politics as part of a wider challenge to state power and nationalism premised on "exclusionary practices and discourses" (Paasi 2011, 16). The border-sited Indigenous, labour, anti-capitalist, anti-war, and pro-refugee activisms in Canadian Niagara offered alternatives that were, in turn, linked up with wider counter-hegemonic movements committed to more equitable forms of human "security," including "no border" projects and visions predicated on the fundamental freedom to move or *not* move in a globalizing world (Munck 2008, 1239; Anderson, Sharma, and Wright 2009, 11).

Conclusion

In this chapter, I have worked to illuminate how border life, bordering, and everyday nationalism in Canadian Niagara engaged with a wider globalization. While "global talk" was less prominent in the interviews and local press, in the analysis I was able to show how border residents in Canadian Niagara did, in fact, understand themselves as global subjects. Examples of this included their discussion of a globalized Canadianness consistent with a dominant trope of Canadian nationalism that portrayed Canadians as globally welcomed (especially relative to Americans). Accounts of global tourism within Canadian Niagara itself, meanwhile, were linked to claims of cosmopolitan multiculturalism – also a key trope of dominant Canadian nationalism (Banting and Kymlicka 2010) – even while evaluative commentary about the relative compatibility of ethnonationally differentiated tourists reinforced dominant constructions of white-settler Canadianness.

Discussion of multiscalar bordering revealed further border resident constructions of Canada and Canadianness as threatened by "minorities" and "multiculturalism" allegedly produced by an overly permeable Canadian border in a globalized world. Meanwhile, the dangers faced by unauthorized global migrants at a restrictive Niagara border were portrayed as self-induced rather than structurally produced, and constructions of allegedly benevolent border enforcement served to reinforce dominant constructions of a benign Canadian border. Experiences and imaginings of filtered bordering and nationalized identity in Canadian Niagara were forged in relationship to Americanness as well as imagined global Others in ways that often reproduced exclusion and inequality, but these coexisted with an array of counter-hegemonic border-related actions and visions, some of which included fundamental challenges to the existence of borders and processes of bordering in an unequal world.

Conclusion

Wilson and Donnan (2012a) write that "borders require continual reinscription and reperformance, on the part of citizens, governments and other groups both within the state and beyond" (19). This case study of border life, bordering, and Canadianness at Niagara illuminates these processes from the ground up in a particular place and over a particular time period. Before offering some concluding thoughts about the study, I first provide a brief update of some of the changes that have shaped the Canada/U.S. border and Canadian Niagara since the interviews were completed in 2004.

Top-down border projects of the post-research period include the 2005 trilateral Canada, U.S., and Mexico Security and Prosperity Partnership, which was replaced by bilateral Beyond the Border initiatives between the U.S. and Mexico and then the U.S. and Canada in 2010 and 2011, respectively (Ayres and MacDonald 2012). The extent to which these developments represented a shift away from, or consolidation of, post-9/11 U.S. unilateralism in North America border making is debated (Hale 2011; Clarkson 2012), as is the related issue of whether these mark greater North American de-bordering or re-bordering. While some see an attempt to "extend the U.S. security perimeter to the continent's external boundaries" (Clarkson 2012, 101), others point out that such efforts are in apparent tension with U.S. protectionism, weakened NAFTA provisions, and increased U.S. border personnel at the Canadian border (Gilbert 2012, 201; Brunet-Jailly 2012, 109).

In terms of the latter, a 2010 U.S. report claimed that while U.S.-bound "inadmissible alien" apprehensions at the Canada/U.S. border were only 1.3 per cent of those at the Mexico-U.S. border (approximately 7,000 versus 541,000), the Canada/U.S. border required increased investment in enforcement due to a higher "terrorist threat" and the fact that only 32 of 4,000 miles of the Canada/U.S. border (excluding the Alaska/Canada section) were at an "acceptable level" of securitization (United States Government Accountability Office 2010,1).

While there are recurrent calls from some U.S. leaders for equally securitized U.S./Mexico and U.S./Canada borders, disproportional U.S. policing of the Mexican versus Canadian border remains, despite increased staffing at both sites. In contrast to the ongoing construction of Mexican migrants as targets of U.S. border enforcement (Alvarez 2012, 548), moreover, Canadians are still rarely described as the targets of U.S. securitization efforts, as the latter tends to construct non-Canadian terrorists crossing the Canada/U.S. border as the putative threat.

Ongoing global displacement and forced migration combined with restrictive North American borders and migration regimes, meanwhile, produce ongoing flows of unauthorized migrants that, in turn, are portrayed by the U.S. (and, to a lesser extent, the Canadian) government as evidence of the need for increasing border enforcement. It is apparent, for example, that by reducing authorized refugee migration to Canada from the U.S., the Safe Third Country Agreement has contributed to an increase in unauthorized U.S.-to-Canada border crossings (Arbel 2013; Arbel and Brenner 2013). In Canada, among other changes to migration management, the 2012 Preventing Human Smugglers from Abusing Canada's Immigration System Act aimed to increase criminalization of such unauthorized flows.

For many Canadian Niagara border residents, a more directly felt top-down border project was the U.S.-initiated Western Hemisphere Travel Initiative (WHTI), first announced in 2005 and implemented 1 June 2009. This initiative first proposed a passport requirement for all entries or re-entries to the U.S. from Canada. The proposal alarmed border elites, including those in Niagara, who pointed to low rates of passport possession: at the time, 23 per cent of Americans and 44 per cent of Canadians ("Mayors Call for Delay in US Passport Plan," *The Standard*, 21 July 2006, A1). The subsequent inclusion of passport alternatives (including an "enhanced" driver's licence or Nexus Card) reflected, in part, the lobbying efforts of an array of Canadian and U.S. "border area chambers of commerce and tourism associations" (Hale 2011, 54–6).

As in the past, Canadian Niagara border life continues to be impacted by fluctuating exchange rates. Through the late 2000s, as the Canadian dollar moved towards parity (achieved in late 2007), U.S.-bound cross-border shopping increased and then remained high, further encouraged by increased spending allowances introduced by the Canadian government mid-2012 (and ongoing weak enforcement by Canadian customs officials of the obligation to pay taxes and duties on same-day U.S.-bought goods).[1]

An example of a local elite response to these outbound flows was a "1 Less Trip" initiative, launched in 2013 by the Greater Niagara Chamber of Commerce. The campaign aimed to convince Canadian residents to take one less trip across the border to benefit the Canadian-side economy, even while a spokesperson for a Canadian-side mall acknowledged that cross-border shopping, dining, and entertainment were part of the culture of Canadian Niagara. Reduced U.S-bound flows, however, became discernable only when the Canadian dollar began to drop again in early 2015 ("Winners of a Low Canadian Dollar in Niagara," *Niagara Falls Review*, 5 February 2015, A2).

The new requirement for WHTI-compliant documents, combined with the period of more favourable exchange rates from 2007 to 2014, did serve to dramatically increase Canadian-side enrolment in Nexus (which, as mentioned, meets the WHTI requirement).[2] In Canadian Niagara, however, Nexus card holders have complained of an onerous application process and "harsh" treatment of those who cross with undeclared goods or transport a passenger without a Nexus card (the latter revealing that Nexus has not eliminated long-standing patterns of cross-border smuggling of unauthorized goods and people) (Binational Economic and Tourism Alliance 2011, 44).[3] It has also been suggested that the new U.S. documentary requirements have had the effect of "offending many Canadians who felt a sense of entitlement to easy access to the United States that was part of the borderland quality of life" (Binational Economic and Tourism Alliance 2011, 41).

Concern continues to be expressed by Niagara elites that WHTI and other securitization initiatives are slowing and reducing regional cross-border commercial and tourist flows despite changes in border infrastructure aimed at expediting these mobilities (including additional Nexus lanes and a 2014 six-month pilot program for pre-clearance of U.S.-bound trucks on the Canadian side of the Peace Bridge).[4] As in the past, the proffered "solution" to the "problem" of slowed and/or declining cross-border flows at Niagara, invokes discourses of official cross-border regionalism that promote working towards a "new binational culture" marked by a "sense of regional connectedness and understanding of the importance of the Niagara border to the North American economy" (Binational Economic and Tourism Alliance 2011, 9).

As central and more regional top-down border projects of both the U.S. and Canada continue to combine with shifting exchange rates to shape border life and border crossings on the ground, the latter also continue to be marked by filtered bordering. The "fast track" option offered by Nexus and other pre-clearance programs continue to expedite some cross-border mobilities, but Indigenous and ethnoracialized border crossers continue to report more troubled border inspection experiences (Simpson 2014; International Civil Liberties Monitoring Group 2010; Canadian Human Rights Commission 2011; "Profiling, Congestion at Issue in Border Forum," *Buffalo News*, 22 September 2015, 21). As indicated, as authorized migration routes are reduced, some global migrant flows, including those of refugees, have also more frequently been forced into dangerous and sometimes fatal unauthorized border-crossing attempts.

When it comes to everyday nationalism, despite the continued discourse of official cross-border regionalism espoused by some elites in Niagara, there is little evidence that this project is being embraced from the "bottom up" in Canadian Niagara. At the continental level, moreover, there is also little indication of official or popular embrace of the idea of North America as either a political or cultural locus of imagined belonging. Canadian nationalism remains strong, and official and more popular expressions continue to be marked by ethnoracialized exclusions and formulations of core and peripheral Canadianness (Winter 2014, 2015).

This study of Canadian border life in Niagara has offered a rich case study aimed at illuminating local experience and perception of top-down border projects, filtered bordering, and nationalized Canadianness. While the project began with a focus on the ostensible de-bordering initiated by top-down free trade agreements, it ended up documenting local experience of forms of filtered re-bordering associated with post-9/11 securitization. The resulting analysis, while distinguishing between the pre- and post-9/11 periods, reveals continuities as well as changes in local border engagements across this chronological divide.

The ways in which Canadian border residents discussed top-down border projects, including official cross-border regionalism, revealed complex forms of engagement that included support and/or acquiescence as well as more critical and/or transgressive positions vis- à-vis state power. The findings point to the importance of supplementing scholarly and more policy-oriented research focused on the border projects of political and economic elites, with greater attention to the ways in which those projects are engaged with on the ground by those living in border regions.

Such a focus, however, should not be simply one of shifting the scale of research but of also broadening the scope of research topics to prioritize issues of concern to everyday border residents. While some border-related research listens to border residents in order to ensure more effective implementation of top-down border projects, this study had the broader goals of exploring and amplifying local perspectives on the border and bordering as part of a broader challenge to the current "democratic deficit" that marks North American border initiatives (Ayres and MacDonald 2012, 18–22; Gilbert 2012, 202). Prioritizing the voices of everyday Canadian border residents, as this study has done, can reveal disjunctures between elite border projects and more everyday experiences and preoccupations thereby challenging claims by the powerful that their border-related projects are congruent with the interests of all.

While committed to the exploration and dissemination of more subjugated border experiences and perceptions, the "bottom-up" research of this study also challenges portrayals of homogenized border culture by highlighting the existence and significance of diversity and inequality among border residents. In particular, the analysis pointed to how, in a context of shifting exchange rates and pre- and post-9/11 securitization, Canadian border residents found themselves differently and unequally positioned in terms of relatively eased or troubled filtered border crossings.

That these unequal positionings had material consequences in border life was clear insofar as those who were more privileged border crossers could maximize the economic, cultural, and social opportunities of a border region. By contrast, those whose crossings were more constrained were less likely to accrue such benefits. Along with the reproduction of material inequalities through mobilized privilege and constraint, other divisions were also reproduced as some Canadian border residents found that their border crossings reinforced a sense of being desirable "safe citizens" within national, binational, and global space. Those who anticipated or experienced more trouble at the border, on the other hand, experienced being deemed inherently "unsafe" on the basis of officially perceived attributes and exclusion from national, binational, and/or global belonging. Filtered bordering in this way reproduced inequality as well as forms of non-belonging in the border region, and these require greater critical attention.

Filtered bordering emerged as significant in terms not only of inequality among Canadian border residents but also of non-local flows. Border resident narratives and the local press reporting on the issue of global migration, and in particular the dangerous and sometimes fatal crossings associated with unauthorized crossings, provided (even as Canadian complicity was downplayed) clear evidence of how bordering at Niagara is linked up with multiscalar inclusions and exclusions that have life-and-death consequences. While national responsibility was deflected in most accounts, such evidence challenges still-dominant constructions of a benign Canadian border in a globalized world.

The project also revealed how the varieties of borderline Canadianness constructed by Canadian border residents articulated with Americanness, global Otherness, and white settlerism. While much of the everyday Canadianness documented here parallels other studies of Canadian nationalism, the ways in which it articulated with regional, classed, and ethnoracialized inequality, as well as its link to tourism and migration, point to the specificity of place-based nationalisms. The "bottom-up" approach illuminates the complexity of everyday Canadian nationalism, including its articulation with unequal multiscalar structures and processes as well as, in some cases, more alternative border activism and visions that challenge such structures and processes. The case study of border life, bordering, and everyday nationalism in Canadian Niagara thereby contributes to a greater understanding of borders, bordering, and nationalism in a globalized world.

Appendix: Interview Schedule

1) Background

While there has been a lot of debate in Canada about the impact of globalization, little is known about the actual experiences of children growing up in a border region like Niagara. We are interested in learning more from you about the everyday experience of growing up close to the border.

It would help us to better understand your childhood experiences if you could tell us a little bit about your history of living in a border community/region, for instance, what age were you? Which years did you live there? Can you also tell us something about your family background in terms of length of residence in the border region? Educational and/or occupational background? Cultural background?

2) Crossing the Border

We now want to focus on your experiences as a child growing up in a border community/region and begin by asking questions about your experience of border crossing.

How often and with whom did you cross the border at Niagara as a younger child? Did this change as you got older? What were some of the reasons that you crossed the border, for instance, shopping, visiting friends or family, tourism, recreation, etc? Did the reasons change as you got older? What happened as you crossed the border, for instance, experiences with border officials? Did these change as you got older? What did it feel like to cross the border when you left Canada and when you came back to Canada? Can you provide any stories about particular crossings that you can still remember?

Do you think that your experiences of border crossing were similar or different to those of other kids and families that you knew? Why did other kids and families cross the border? What stories do you recall hearing from other children about this experience?

As a child, were you aware that some people might have problems crossing the border, for instance, friends and neighbours, tourists, immigrants, refugees? As a child, were you aware of the risks that some people took in border crossing, for instance, being turned back, arrested, deported, injured, or killed in the crossing process? What factors do you think make it easy or difficult for people to cross today?

3) Other Borders

What, if any, other borders did you cross as a child? Did you have family or friends in other countries whom you visited or who visited you? Did you communicate with family and/or friends in other countries via phone, letters, Internet? Did many kids that you knew have family or friends in other countries? Can you provide any examples?

4) Living Near the Border

Apart from the experience of border crossing itself, in what other ways do you think that living near the border affected your life as a younger/older child? Did you interact with non-Canadians on the Canadian side of the border? Did you have any work experiences that were linked to living in a border community? How did living in a border community affect the lives of other kids you knew? Can you provide any examples?

5) Childhood and Identity

Can you recall for us how you tended to identify yourself as a child? Was it for example, important to be from Niagara and/or Canada? Do you think these identities were important for other kids that you knew? What other identities were important to you? Other kids you knew? Can you provide any examples? Has your own sense of identity changed over time?

6) National Identity

People have many ideas about what Canada and being Canadian is like. What were your impressions of Canada and Canadian identity as a child? Can you provide any examples? People have many ideas about what America and being American is like. What were your impressions of America and Americans as a child? Can you provide any examples? Are there any other countries that you know a lot about? Could you describe your impressions of that country and those living there? Some people think that being Canadian, American, or another nationality doesn't matter so much now because some things like the economy, the media and/or environmental issues are affecting the whole world, not just one or some countries. Do you think that most kids in Niagara feel more connected to the whole world than part of a particular city/region/country? Do you think that even if you are part of a larger world it is still important to be Canadian or American or another nationality? Why or why not? Do you think that being part of a larger world has an impact on the lives of children and youth?

7) Globalization

There has been a lot of debate in recent years about globalization and its effects. One straightforward definition of globalization would be an increased movement of people, goods, money, and/or ideas across national borders. Is this increased movement of people, goods, money, or ideas something that you have been aware of growing up in a border region?

Some argue that not only is there increased movement across borders but also national borders are changing. Has your lived experience of the border changed over time?

It has been suggested that globalization is linked to changes in the ways that governments provide services to citizens, including children. Can you comment on recent changes in the areas of schools, health care, youth employment, and/or child and youth programs in the Niagara border region? Do you think that your own life has been affected by these changes? If so, in what ways? Can you provide some examples? What about the lives of others you know? Can you provide examples?

8) Other Issues

Are there any other aspects of being a "border kid" that you think are important to this study?

Notes

Introduction

1 For examples, see Revie (2003), Nicks and Grant (2010), and Jackson (2003).
2 See Newman (2011), Wastl-Walter (2011), Wilson and Donnan (2012b), and Popescu (2012).
3 For examples, see Berdahl (1999), Donnan and Wilson (1999, 2010), Vila (2000, 2003, 2005), and Wilson and Donnan (1998, 2012b).
4 See Neumayer (2006, 2010), Goldring, Berinstein, and Bernhard (2009), Johnson et al. (2011), Mountz (2011), Mountz and Hiemstra (2012), Coleman (2012), Satzewich (2015).
5 For examples, see the movie *The River*, television series such as *The Border* and *Border Security,* Roberts and Stirrup (2013), and Laxer (2003).

1. Bordering Canada at Niagara

1 See also Bellfy (2013), McManus (2005), Jameson and Mouat (2006), McDougall and Philips (2012, 180), Hele (2008), Singleton (2009), Manore (2011), Boos and McLawsen (2013), Feghali (2013), and Graybill and Johnson (2010).
2 See Hier and Greenberg (2002), Mountz (2010), Park (2010), and Kazimi (2012). For the more limited anti-migrant discourses and actions at the eastern coastal border, see Abella and Troper (1982), Mannik (2013, 2014), and Anderson (2013, 145–6).
3 See Byers (2009), Heininen and Zebich-Knos (2011, 198–9), and Nicol (2011).
4 Anti-Asian measures in Canada, historically coordinated with the U.S. and Britain, were enacted at both coastal and land borders in North

America. See Sohi (2011) and Price (2011, 2013). Within the first decade of the 1900s, these provisions combined with other exclusionary measures (e.g., "continuous journey" provisions) to dramatically curtail legal Asian migration to Canada and the U.S. See Ngai (2004, 18), Anderson (2013, 82), and Sadowski-Smith (2014).

5 For more details on the Borderlands Project, see Konrad and Nicol (2008, xv-xvi, 46–7).

6 As an example of the former, see the many reports of the Border Policy Research Institute (http://www.wwu.edu/bpri). The smaller studies of Ziolkowski (2006) and Paulus and Asgary (2010) surveyed U.S. border residents at Niagara and Canadian border residents living near the Ontario/Michigan Blue Water Bridge, respectively. The Paulus and Asgary (2010) study found that Canadian border residents living at the Blue Water Bridge did not perceive a security threat and "tended to not believe there was a direct benefit to the local community from border security enhancements" (20). A spring 2005 survey of U.S. residents living on Grand Island in the Niagara River found a gendered difference in response to questions about border safety, with 46 per cent of female but only about one-quarter of male respondents describing the border to be "very or somewhat dangerous" (Ziolkowski 2006, 8). The majority however, thought the police presence was adequate, and some expressed concern about the cost of border security (10). A 2012 restudy conducted in response to a request from the local U.S. Border Patrol found that residents viewed the Patrol as "overbearing" ("Border Police Use Study to Improve Image," *The Stylus*, 19 November 2013).

7 Scholarly work on the impact of Niagara bordering on Indigenous lives remains limited. For aspects of this, however, see Williamson and MacDonald (1998), Jackson (2003), Grinde (2002), Dirmeitis (2012), and McDougall and Philips (2012). Peace Bridge expansion led to archaeological excavation in 1990s (following earlier work in the 1960s), and in his forward to a book about these excavations, the executive director of the Fort Erie Native Friendship Centre shared his deep ambivalence about the project (Williamson and MacDonald 1998, ix). In the course of proposals for a new or expanded bridge, a spokesperson from the Fort Erie Chapter of Native Women was reported as advocating a Native Spiritual Memorial Garden and Educational Resource Centre dedicated to "those who survived or didn't return home from residential schools" ("Native Group Seeks Role in Gateway," *Niagara Falls Review*, 18 October 1999, A5). See also "Fort Erie Powwow Rekindles Traditions," *Niagara Falls Review*, 30 July 1999, B1; "Bridge Site

Rich with History," *Niagara Falls Review*, 7 August 1999, A1; and "U.S. Politician Backs Twinned Peace Bridge," *Niagara Falls Review*, 23 February 2000, A1. In 2006, a small archaeological display and interpretive centre opened at the newly expanded Peace Bridge, showcasing Indigenous artefacts uncovered during the excavations as well as contemporary "Native artwork." The Peace Bridge website described the exhibit as a collaboration among the Town of Fort Erie, the Fort Erie Native Friendship Centre, Fort Erie Museum Services, and the Buffalo and Fort Erie Public Bridge Authority. See also Dubinsky (1999, 55–83) on the history of the presence and representations of Indigenous peoples in Niagara tourism since the nineteenth century, and Bredin (2010, 63) on the Niagara Falls, Ontario, "Indian Village" tourist attraction of the 1960s.

8 For local ceremonies marking the bicentenary, see "Customs Set to Mark Milestone," *Niagara Falls Review*, 30 April 2001, A2; and "Minister Praises Work of Customs Officials," *Niagara Falls Review*, 26 May 2001, A3.

9 For a discussion of fugitive slaves, Great Lakes bordering, and memorializing, see Wigmore (2011), Faires (2006, 2013), and Chambers and Siener (2012). There were significant black communities in St. Catharines, Niagara-on-the-Lake, Fort Erie, and Drummondville in the nineteenth century, and Fort Erie was the site of the 1905 meeting of the Niagara Movement that would later become the U.S. National Association for the Advancement of Colored People (Walcott 2003, 32; Jones 2011). The Emancipation Day ceremonies (including picnics) in St. Catharines were reportedly among the largest in Ontario from the late 1920s through to the 1950s (Jackson 2003, 155–7).

10 Despite Canada's anti-Chinese legislation, Siener (2008a) writes that these "somewhat less restrictive" provisions led some Chinese to transit through Canada to the U.S. (38). Siener does not refer specifically to Canada-bound human smuggling at Niagara, but Anderson (2013) refers to some U.S.-to-Canada flows (113).

11 While most Buffalo residents opposed Prohibition, the U.S. government was apparently "prepared to sacrifice the peaceful border in favor of its Prohibition enforcement goals" (Siener 2008b, 441). After the U.S. government threatened to place 10,000 armed agents on the border in 1930, the Canadian government finally acceded to a U.S. demand for a ban on liquor exports (Siener 2008b, 442–5).

12 Klug (2010) also documents how, at the Detroit/Windsor crossing, such pressures resulted in early innovations including the institution of primary and secondary inspection, experiments with pre-inspection, "land border crossing cards" and designated lanes for some crossers (402–3).

13 Recruitment materials invited young adults who had spent a substantial portion of their childhood (e.g., five years) in a Niagara border community to participate in a 45- to 60-minute interview on experiences of border crossing and border living (see the appendix). While the first set of interviewees received no compensation, in 2004 I added an honorarium of $20.00 for each participant. Those who responded to recruitment efforts were provided with a copy of the interview schedule prior to the interview and had the opportunity to review and revise their transcript before it was included in the database. Only a few interviewees reviewed their transcript, and only minor revisions were requested. Graduate and undergraduate students transcribed the interviews, while I conducted the analysis by reviewing transcripts to identify recurring topics and themes.
14 I gathered approximately 300 articles from 1999, 170 from 2000, 200 from 2001, 100 from 2002, 200 from 2003, and 200 from 2004.
15 For the younger group, born between 1972 and 1985, the inception of the Canada-U.S. 1989 Free Trade Agreement coincided with early childhood to late teen years (4 to 17 years), and the 1994 adoption of NAFTA coincided with late childhood to early adulthood (9 to 22 years). At the time of 9/11, they were in their late teens to adulthood (16 to 29 years of age).
16 Some had left the border communities of their childhood and teen years to take up residence close to the regional university that was a twenty-minute drive from the border.
17 I describe as "upper-middle class" those with at least one parent with university education or an occupation that required such a credential. Those described as "lower-middle class" had at least one parent with some form of post-secondary education or training or occupation requiring such education or training. Those labelled as "working class" had parents with partial college, high school, or partial high school education or corresponding occupations. The largely university-based recruitment in a region with below-provincial-average university attainment meant that the pool was more privileged than the regional population. Some of the older interviewees occupied different class positions than those they had experienced as young people.
18 The lack of self-racialization by many of the other interviewees is consistent with the insights of those who note that "'whiteness' is apparently difficult for white people to name," despite the ways in which this racialized identity "continuously shapes … experiences, practices and views of the self and others" (Frankenberg 1993, 228).
19 I did not ask directly about citizenship, so this may not be accurate. The number with dual Canadian/U.S. citizenship may be indicative of

the border region or reflect self-selection for this border-related project. The 2006 census provided a figure of 1.58 per cent U.S.-born for the St. Catharines-Niagara Census Metropolitan Area – the third highest after the border regions of Windsor at 2.05 and Victoria, BC, at 1.85 (Hardwick and Smith 2012, 302).

2. Growing Up at the Borderline Pre-9/11

1 Stephens (1995), Scheper-Hughes and Sargent (1998), and Hess and Shandy (2008). For children, youth, and border crossings and identities, see Lask (1994) Hipfl, Bister, and Strohmaier (2003), Dunkley (2004), Bejarano (2010), Christou and Spyrou (2012), and Spyrou and Christou (2014).

2 The introduction of a Canadian Goods and Services Tax in 1991 also encouraged cross-border shopping at the end of this period (Traister 2002). Chuck, who lived on the U.S. side during high school, reported that during this period, "Canadians were frowned upon … [and] got a bad rap because they would shop and … leave their boxes in the parking lots and wear the [undeclared] clothes across."

3 A second U.S. airport, the Niagara Falls International Airport, which is also very close to the border, was expanded in 2009 and is popular with Canadians.

4 In 1985, only ten U.S. retailers were operating in Canada, but by 2005 there were about 100 ("Shoppers Making Fewer Trips to U.S.," *Globe and Mail*, 3 March 2005). Dual citizen June, who grew up on the other side, recalled shopping in Canada for butter tarts, a particular kind of coffee, peameal bacon, and clothes that were "a bit different" and therefore considered "pretty cool."

3. Experiencing 9/11 and Post-9/11 Securitization at the Borderline

1 The CANPASS pre-clearance system was initiated by the 1995 Canada–United States Accord on Our Shared Border and introduced to Niagara when the Whirlpool Bridge became (controversially) an exclusive CANPASS crossing in 1998.

2 See also "Bomb Scares Shut Peace, Q-L Bridges: U.S. Customs Investigating Reports," *Niagara Falls Review*, 13 September 2001, A1.

3 See "'Quiet Anger' Grips America: Community Pitches In At 'Trucker Town,'" *Niagara Falls Review*, 14 September 2001, A1; "City Opens Heart to Stranded Truckers," *Niagara Falls Review*, 15 September 2001, A5.

4 The suggestion that 9/11 terrorists entered from Canada persists in the U.S., while the Millennium Bomber case – which better supports a link between the U.S./Canada border and U.S.-bound terrorists – is rarely referenced (Salter and Piche 2011, 946).
5 For reflections on post-9/11 developments affecting the local Muslim and South Asian community in Niagara and nearby Hamilton, where there were also attacks on a mosque and Hindu Temple, see "Muslims Faced Up To 'Fear of Other' Post-9/11," *Welland Tribune*, 8 September 2011, online edition (http://www.wellandtribune.ca/2011/09/08/muslims-faced-up-to-fear-of-other-post-911), and "Three Men Charged in 2001 Hindu Temple Arson," *Globe and Mail*, 8 November 2013, A8.
6 Key earlier agreements were the 1999 Canada-U.S. Partnership Forum and the 1995 Canada–United States Accord on Our Shared Border.
7 See also "Federal Budget Beefs Up Border Security," *Niagara Falls Review*, 11 December 2001, A1.
8 See Binational Economic and Tourism Alliance (2011, 86–7) for a listing of changes to the physical infrastructure of the bridges during the decade following 9/11.
9 See "New Bridge Not Dream Come True," *Niagara Falls Review*, 11 May 2002, A4, and "Bridge Conversion Could be Good for City," *Niagara Falls Review*, 14 November 2002, A4.
10 See "Bridge Delays May Affect Your Health," *Niagara Falls Review*, 13 November 2003, A3, and "PBA Won't Support New Bridge: Focusing On Moving Plaza," *Niagara Falls Review*, 5 December 2003, A1.

4. Filtered Bordering and Borderline Lives

1 According to the *Niagara Falls Review* reporting, the first protest was focused on a demand for the release of "political prisoners" (Mumia Abu-Jamal, Leonard Peltier, and Dudley George). The second protest (described as also supported by the Workers World Party and the Canadian Auto Workers Local 199) was reportedly focused on highlighting hunting and fishing rights, conditions on reservations, the prisoners named above, and the need for both countries to respect border-crossing rights under the 1784 Jay Treaty. Some members of the Haudenosaunee Confederacy would have used a Haudenosaunee identity card and passport for border crossing in this period. This documentation is unevenly recognized by the U.S. government and has received even less support from the Canadian state (Dirmeitis 2012).

2 The 1952 Immigration Act included homosexuals in the "prohibited classes" for entry into both Canada and the U.S. (Pratt 2005, 77; see also Luibhéid 2002).

3 The first 1875 U.S. federal restrictions excluded prostitutes, with the goal of denying entry to Chinese women (Ngai 2004, 58–9). See Pickering and Ham (2014) on the dynamics of border control and sex work at another border. For a discussion of gendered borders more generally, see Yuval-Davis and Stoetzler (2002).

4 Tanovich (2006) defines racial profiling in policing as "heightened scrutiny" "based solely or in part on race, ethnicity, Aboriginality, place of origin, ancestry, or religion or on stereotypes associated with any of these factors rather than on objectively reasonable grounds to suspect that the individual is implicated in criminal activity" (13). The interviewee focus on blackness for the pre-9/11 period underscores Wortley and Tanner's (2005) caution that broad terms such as "visible minority" (or "non-white") in research on racial profiling can "mask important racial differences in both experience and behaviour and ultimately hinder the identification of racism in Canada" (588; see also Tator and Henry 2006).

5 For racial profiling and interracial relationships in Canada, see Deliovsky (2010) and Kitossa and Deliovsky (2010).

6 There are striking parallels between this response and those of South Asian Muslims in Montreal (Jamil and Rousseau 2012).

5. Everyday Nationalism at the Borderline

1 Politicians on the Canadian side have long lobbied for a new mid-peninsula highway to facilitate continental commercial flows ("Linking of Trade Routes Beneficial to Canada, U.S.," *Niagara Falls Review*, 23 June 2000, A6), and the Continental One trade corridor linking Toronto to Miami through Niagara was promoted by U.S.-side elites as a "vital link for the eastern seaboard in the free-trade era economy" ("Pataki Pitches Federal Cooperation to Prop Up Border," *Niagara Falls Review*, 27 June 2001, A3).

2 Eagles made this point in relation to the broader region that encompasses Toronto as well as Western New York, noting that "the imbalance created by the size and power of the United States in comparison to Canada is reversed … [as] Toronto and Southern Ontario constitute the demographic and socioeconomic core of Canada, and in this respect this Canadian region stands in stark contrast to the peripheral status of western New York" (Eagles 2010, 381).

3 The role of schools in the production of nationalized identities was also clear when Chuck, who spent high school in the U.S., reported that while British victories of the War of 1812 were highlighted in the Canadian curriculum, American triumphs were emphasized in the U.S. (Heyman 2012, 54; Rippberger and Staudt 2003).

4 Donnan and Wilson (1999, 91–5, 143–8); Vila (2005); Campbell (2007). Dubinsky's (1999) history of the heterosexualized honeymoon industry at Niagara Falls does not address the border itself.

5 These American grandmothers had ensured that their children had dual citizenship by giving birth on the U.S. side, despite having permanent residence in Canada. A tendency toward patrilocal residence in cross-border marriages could also be producing the gendered pattern in my Canadian side interview pool.

6 That Amanda's anti-Americanism was mediated by class was suggested by her much more positive account of working at "very rich" U.S. summer camp on the other side of the river, where the children came from "pretty affluent" households.

7 See E. Helleiner (2006) on Canadian currency and nationalism.

8 While there were few interviewee references to Quebec, French-Canadian identity was occasionally mentioned, and some interviewees identified Francophone parents. While Canadian Niagara contains a significant francophone population, these comments about the U.S. visitors appeared to construct an exclusively anglophone Canadianness (for more on the role of Quebec in English-Canadian nationalism, see Winter 2007).

9 For documentation of employment conditions in the Canadian Niagara hotel sector, see Hickey (2008).

10 Hardwick and Mansfield's (2009) study of U.S.-born migrants to British Columbia also found "strong and consistent … adoption of a distinctively non-American Canadian identity" rather than a hybridity or binationalism (396). They attribute what they describe as rapid "Canadianization" to the fact that these are voluntary migrants (often political dissenters rather than refugees) marked by anglophone whiteness and class privilege. Hardwick's (2010) study of U.S.-born migrants in Toronto, Vancouver, and Halifax (which includes references to anti-Americanism experienced by elite migrants in Toronto in the 1970s), however, cautions that those who arrive as privileged (often corporate) economic migrants may be more attached to U.S. citizenship and identity than those who arrive as political dissenters. See also Hardwick and Smith (2012).

6. Bordering Globalization at the Borderline

1 Some had travelled to European countries as tourists or to visit relatives. Others had vacationed in Bermuda, the Caribbean, or Mexico. One had been an English-language worker in Asia, while others had engaged in missionary or pilgrimage travel (to Grenada, Mexico, Ukraine, Israel, and India).

2 For discussion in the context of Serbia, claims that global mobility may produce "the right kind of citizens" while restricted travel may lead to problematized forms of nationalism, see Greenberg (2011, 96).

3 See "U.S. Border Patrol Arrests Seven," *Niagara Falls Review*, 8 April 2003, A7; "Child Slips Across Border: Seven-Year-Old Crosses Whirlpool Bridge, Found in NFNY," *Niagara Falls Review*, 12 May 2003, A1; "Teen Crosses Bridge Undetected," *Niagara Falls Review*, 3 June 2003, A3.

4 See "Women Risk Death in Smuggling Attempt: 4 Wedged Under Truck at Canada-U.S. Border," *Globe and Mail*, 1 February 1999, A1; "'Death-Defying' Smuggling Attempt Is Thwarted at the Border," *New York Times*, 1 February 1999, B6.

5 See "Border Jumpers Might Have Paid $40,000 Each," *Niagara Falls Review*, 10 February 1999, B1; "Six Chinese Arrested in Smuggling Bid Quickly Escalating Trade in Illegal Migrants Is 'Major Problem,' U.S. Border Officials Say," *Globe and Mail*, 10 February 1999, A3.

6 See also "Grim Discovery at Peace Bridge: Body of Stowaway Wrapped around Rear Axle of Bus," *Niagara Falls Review*, 3 January 2001, A1.

7 See "Nine Illegals Captured on Raft," *Niagara Falls Review*, 7 June 2004, A3; "A Canadian Gate Where Illegal Immigrants Knock," *New York Times*, 15 June 2004, B5.

8 The extent of these crossings in either direction is difficult to establish. The major data source records only apprehensions of those attempting entries between (not at) ports of entry. This figure, publicly available for 2006 was 1,113 U.S.-bound and 79 Canada-bound for the entire border (Canada–United States Integrated Border Enforcement Team Threat Assessment 2007). The other publicly available figures for 2008 and 2009 are 819 and 779 U.S.-bound apprehensions and 952 and 842 Canada-bound apprehensions, respectively ("Cross-border Human Smuggling Tilted toward Canada, Says Border Intelligence Report," *Winnipeg Free Press*, 17 September 2010; Canada–United States Integrated Border Enforcement Team Threat Assessment 2010).

9 For federal public sector workers, Bell employees, and customs officers, see "PSAC Pickets at Bridge," *Niagara Falls Review*, 13 March 1999, C1;

"Bell Protestors Warned of Bomb Threat," *Niagara Falls Review*, 1 May 1999, A1; "Customs Officers Call Off Picket," *Niagara Falls Review*, 9 May 2002, A2). Canadian Members of the Industrial, Wood and Allied Workers also used the border to protest U.S. duty on Canadian softwood lumber by handing out leaflets to drivers ("Drivers Targeted Over Softwood Lumber Dispute: Part of Campaign to Cover 50 Border Crossings," *Niagara Falls Review*, 27 May 2002, A3). See also "Tense Moments at Peace Bridge: Traffic Chaos as Demonstrators Protest Free Trade Agreement," *Niagara Falls Review*, 23 April 2001, A1; "Refugees' Rights Protest Held at Border," *Niagara Falls Review*, 10 May 2003, A3.

Conclusion

1 "Cross-border Shopping a $20B Economic Drain," *Canadian Press*, 17 May 2012; "Border Guards Too Lax on Shoppers: Study," *Globe and Mail*, 24 March 2014, A4.

2 Nexus had 265,000 enrollees at the Canada/U.S. border by January 2009 (approximately 70 per cent of these were Canadian).

3 See also, "'Low Risk' Travelers Caught Smuggling Beer, Jewellery into Canada," *Thestar.com.*, 17 July 2011; "Nexus Members Nabbed at the Border; 'Low Risk' Travellers Attempted to Bring Unauthorized Passengers into the Country, Documents Show," *Ottawa Citizen*, 15 January 2012, A1.

4 The 2011 Binational Economic and Tourism Alliance–commissioned report suggested that Niagara border wait times were, however, shorter than generally perceived, and that despite a drop in U.S. visitors to Canadian Niagara following 9/11, commercial crossings actually increased between 2001 and 2008. The overall declines – of concern to local elites – were attributed less to border securitization than to exchange rate fluctuations, the U.S. 2008 recession, and changing patterns of tourism and travel (Binational Economic and Tourism Alliance 2011, 68). Local reporting from late 2014, however, highlighted ongoing delays ("Back-ups at Border Crossings Are Longer, More Frequent This Summer," *Buffalo News*, 29 August 2014).

References

Abella, Irving, and Harold Troper. 1982. *None Is Too Many: Canada and the Jews of Europe, 1933–1948*. Toronto: Lester and Orpen Dennys.

Alvarez, Robert R. 2012. "Reconceptualizing the Space of the Mexico-US Borderline." In *A Companion to Border Studies*, edited by Thomas Wilson and Hastings Donnan, 538–56. Malden, MA: Wiley-Blackwell. http://dx.doi.org/10.1002/9781118255223.ch31

Anderson, Bridget, Nandita Sharma, and Cynthia Wright. 2009. "Editorial: Why No Borders?" *Refuge: Canada's Periodical on Refugees* 6 (2): 5–18.

Anderson, Christopher. 2013. *Canadian Liberalism and the Politics of Border Control 1867–1967*. Vancouver: University of British Columbia Press.

Andreas, Peter, and Thomas J. Biersteker. eds. 2003. *The Rebordering of North America: Integration and Exclusion in a New Security Context*. New York: Routledge.

Andreas, Peter. 2005. "The Mexicanization of the US-Canada Border: Asymmetric Interdependence in a Changing Security Context." *International Journal* 60 (2): 449–62. http://dx.doi.org/10.2307/40204302

Andrucki, Max. 2010. "The Visa Whiteness Machine: Transnational Motility in Post-apartheid South Africa." *Ethnicities* 10 (3): 358–70. http://dx.doi.org/10.1177/1468796810372301

Anzaldua, Gloria. 1999. *Borderlands: La Frontera*. 2nd ed. San Francisco: Aunt Lute Books.

Arbel, Efrat. 2013. "Shifting Borders and the Boundaries of Rights: Examining the Safe Third Country Agreement between Canada and the United States." *International Journal of Refugee Law* 25 (1): 65–86. http://dx.doi.org/10.1093/ijrl/eet002

Arbel, Efrat, and Alletta Brenner. 2013. *Bordering on Failure: Canada-U.S. Border Policy and the Politics of Refugee Exclusion*. Report prepared for the Harvard

Immigration and Refugee Law Clinical Program. Cambridge, MA: Harvard Law School.

Ayres, Jeffrey, and Laura MacDonald. 2012. "Introduction: North America in Question." In *North America in Question: Regional Integration in an Era of Economic Turbulence*, edited by Jeffrey Ayres and Laura MacDonald, 3–30. Toronto: University of Toronto Press.

Banting, Keith, and Will Kymlicka. 2010. "Canadian Multiculturalism: Global Anxieties and Local Debates." *British Journal of Canadian Studies* 23 (1): 43–72. http://dx.doi.org/10.3828/bjcs.2010.3

Bejarano, Cynthia. 2010. "Border Rootedness as Transformative Resistance: Youth Overcoming Violence and Inspection in a US-Mexico Border Region." *Children's Geographies* 8 (4): 391–9. http://dx.doi.org/10.1080/14733285.2010.511004

Bellfy, Phil. 2013. "The Anishnaabeg of Bawating: Indigenous People Look at the Canada-US Border." In *Beyond the Border: Tensions across the Forty-Nine Parallel in the Great Plains and Prairies*, edited by Kyle Conway and Timothy Pasch, 199–222. Montreal: McGill-Queen's University Press.

Berdahl, Daphne. 1999. *Where the World Ended: Reunification and Identity in the German Borderland*. Berkeley: University of California Press.

Bhandar, Davina. 2008. "Resistance, Detainment, Asylum: The Onto-Political Limits of Border Crossing in North America." In *War, Citizenship, Territory*, edited by Deborah Cowen and Emily Gilbert, 281–302. New York: Routledge.

Binational Economic and Tourism Alliance. 2011. *Niagara Border Study: 10 Years after 9/11*. Niagara Falls, ON: Binational Economic and Tourism Alliance.

Boos, Greg, and Greg McLawsen. 2013. *American Indians Born in Canada and the Right of Free Access to the United States*. Research Report No. 20. Bellingham: Border Policy Research Institute. Western Washington University.

Borrows, John. 1997. "Wampum at Niagara: The Royal Proclamation, Canadian Legal History and Self-government." In *Aboriginal and Treaty Rights in Canada: Essays in Law, Equity and Respect for Difference*, edited by Michael Asch, 155–72. Vancouver: University of British Columbia Press.

Bredin, Marian. 2010. "Niagara Falls Indian Village: Popular Productions of Cultural Difference." In *Covering Niagara: Studies in Local Popular Culture*, edited by Joan Nicks and Barry Keith Grant, 45–65. Waterloo, ON: Wilfrid Laurier University Press.

Browne, Simone. 2005. "Getting Carded: Border Control and the Politics of Canada's Permanent Resident Card." *Citizenship Studies* 9 (4): 423–38. http://dx.doi.org/10.1080/13621020500211420

Brunet-Jailly, Emmanuel. 2007. *Borderlands: Comparing Border Security in North America and Europe*. Ottawa: University of Ottawa Press.

Brunet-Jailly, Emmanuel. 2012. "Securing Borders in Europe and North America." In *A Companion to Border Studies*, edited by Thomas Wilson and Hastings Donnan, 100–18. Malden, MA: Wiley-Blackwell. http://dx.doi.org/10.1002/9781118255223.ch6

Bukowczyk, John J., Nora Faires, David R. Smith, and Randy Williams Widdis. 2005. *Permeable Border: The Great Lakes Basin as Transnational Region, 1650–1990*. Pittsburgh: University of Pittsburgh Press.

Byers, Michael. 2009. *Who Owns the Arctic? Understanding Sovereignty Disputes in the North*. Vancouver: Douglas and McIntyre.

Campbell, Howard. 2007. "Cultural Seduction: American Men, Mexican Women, Cross-Border Attraction." *Critique of Anthropology* 27 (3): 261–83. http://dx.doi.org/10.1177/0308275X07080356

Canada–United States Integrated Border Enforcement Team Threat Assessment. 2007. Accessed 26 March 2014, http://www.rcmp-grc.gc.ca/ibet-eipf/reports-rapports/threat-menace-ass-eva-eng.htm (site discontinued).

Canada–United States Integrated Border Enforcement Team Threat Assessment. 2010. Accessed 26 March 2014, http://www.rcmp-grc.gc.ca/ibet-eipf/reports-rapports/2010-threat-menace-eng.htm (site discontinued).

Canadian Human Rights Commission. 2011. *Human Rights Accountability in National Security Practices: A Special Report to Parliament*. Ottawa: Canadian Human Rights Commission.

Carroll, Francis M. 2001. *A Good and Wise Measure: The Search for the Canadian-American Boundary, 1783–1842*. Toronto: University of Toronto Press.

Chalfin, Brenda. 2008. "Sovereigns and Citizens in Close Encounter: Airport Anthropology and Customs Regimes in Neoliberal Ghana." *American Ethnologist* 35 (4): 519–38. http://dx.doi.org/10.1111/j.1548-1425.2008.00096.x

Chambers, Thomas A., and William H. Siener. 2012. "Harriet Tubman, the Underground Railroad, and the Bridges at Niagara Falls." *Afro-Americans in New York Life and History* 36 (1): 34–48.

Chavez, Leo R. 2007. "The Condition of Illegality." *International Migration* 45 (3): 192–6. http://dx.doi.org/10.1111/j.1468-2435.2007.00416.x

Christou, Miranda, and Spyros Spyrou. 2012. "Border Encounters: How Children Navigate Space and Otherness in an Ethnically Divided Society." *Childhood: A Global Journal of Child Research* 19 (3): 302–16. http://dx.doi.org/10.1177/0907568212443270

Clarkson, Stephen. 2006. "Smart Borders and the Rise of Bilateralism: The Constrained Hegemonification of North America after September 11." *International Journal* 61 (3): 588–609. http://dx.doi.org/10.2307/40204192

Clarkson, Stephen. 2012. "Continental Governance, Post-Crisis: Where is North America Going?" In *North America in Question: Regional Integration in an Era of Economic Turbulence*, edited by Jeffrey Ayres and Laura MacDonald, 85–110. Toronto: University of Toronto Press.

Coleman, Matthew. 2005. "U.S. Statecraft and the U.S.-Mexico Border as Security/Economy Nexus." *Political Geography* 24 (2): 185–209. http://dx.doi.org/10.1016/j.polgeo.2004.09.016

Coleman, Matthew. 2012. "From Border Policing to Internal Immigration Control in the United States." In *A Companion to Border Studies*, edited by Thomas Wilson and Hastings Donnan, 419–37. Malden, MA: Wiley-Blackwell. http://dx.doi.org/10.1002/9781118255223.ch24

Cooper, Afua. 2009. "Acts of Resistance: Black Men and Women Engage Slavery in Upper Canada 1793-1803." In *The Politics of Race in Canada*, edited by Maria Wallis and Augie Fleras, 239–49. Don Mills: Oxford University Press.

Cunningham, Hilary, and Josiah McC. Heyman. 2004. "Introduction: Mobilities and Enclosures at Borders." *Identities: Global Studies in Culture and Power* 11 (3): 289–302. http://dx.doi.org/10.1080/10702890490493509

De Genova, Nicholas. 2002. "Migrant 'Illegality' and Deportability in Everyday Life." *Annual Review of Anthropology* 31 (1): 419–47. http://dx.doi.org/10.1146/annurev.anthro.31.040402.085432

Deliovsky, Katerina. 2010. *White Femininity: Race, Gender and Power*. Halifax: Fernwood.

Dhamoon, Rita, and Yasmeen Abu-Laban. 2009. "Dangerous (Internal) Foreigners and Nation-Building: The Case of Canada." *International Political Science Review* 30 (2): 163–83. http://dx.doi.org/10.1177/0192512109102435

Dirmeitis, Mary. 2012. "For Decades, The Haudenosaunee Have Protested a Border They Didn't Draw." *This Magazine*, 26 January.

Doevenspeck, Martin. 2011. "Constructing the Border from Below: Narratives from the Congolese-Rwandan State Boundary." *Political Geography* 30 (3): 129–42. http://dx.doi.org/10.1016/j.polgeo.2011.03.003

Donnan, Hastings, and Thomas Wilson. 1999. *Borders: Frontiers of Identity, Nation and State*. Oxford: Berg.

Donnan, Hastings, and Thomas Wilson. 2010. *Borderlands: Ethnographic Approaches to Security, Power and Identity*. Lanham, MD: University Press of America.

Dubinsky, Karen. 1999. *The Second Greatest Disappointment: Honeymooning and Tourism at Niagara Falls*. Toronto: Between the Lines.

Dunkley, Cheryl Morse. 2004. "Risky Geographies: Teens, Gender and Rural Landscape in North America." *Gender, Place and Culture: A Journal of Feminist Geography* 11 (4): 559–79. http://dx.doi.org/10.1080/0966369042000307004

Eagles, Munroe. 2010. "Organizing Across the Canada-US Border: Binational Institutions in the Niagara Region." *American Review of Canadian Studies* 40 (3): 379–94. http://dx.doi.org/10.1080/02722011.2010.496904

Easson, Gordon. 1999. "Sounds of the Golden Horseshoe: Canadian-American Differences at the Niagara Border." *Toronto Working Papers in Linguistics* 17: 99–110.

Ennamorato, Michael. 2004. "Travel Intentions Study Report: Wave 3 (May 2004)." Report Presented to the Ontario Ministry of Tourism and Recreation. Accessed 26 March 2014, http://www.mtc.gov.on.ca/en/research/travel_intentions/May%202004_Report.pdf

Ennamorato, Michael. 2005. "Travel Intentions Study Initial Results: Wave 6 (May 2005). Report presented to the Ontario Ministry of Tourism and Recreation. Accessed 26 March 2014, http://www.mtc.gov.on.ca/en/research/travel_intentions/May%202005_Initial%20results.pdf

Faires, Nora. 2006. "Going Across the River: Black Canadians and Detroit before the Great Migration." *Citizenship Studies* 10 (1): 117–34. http://dx.doi.org/10.1080/13621020500525993

Faires, Nora. 2013. "Across the Border to Freedom: The International Underground Railroad Memorial and the Meanings of Migration." *American Journal of Ethnic History* 32 (2): 38-67. http://dx.doi.org/10.5406/jamerethnhist.32.2.0038

Feghali, Zalfa. 2013. "Border Studies and Indigenous Peoples: Reconsidering Our Approach." In *Beyond the Border: Tensions across the Forty-Ninth Parallel in the Great Plains and Prairies*, edited by Kyle Conway and Timothy Pasch, 153–69. Montreal: McGill-Queen's University Press.

Ford, Michelle, and Lenore Lyons. 2012. "Labor Migration, Trafficking and Border Controls." In *A Companion to Border Studies*, edited by Thomas Wilson and Hastings Donnan, 438–54. Malden, MA: Wiley-Blackwell. http://dx.doi.org/10.1002/9781118255223.ch25

Fox, Jon, and Cynthia Miller-Idriss. 2008. "Everyday Nationhood." *Ethnicities* 8 (4): 536–63. http://dx.doi.org/10.1177/1468796808088925

Frankenberg, Ruth. 1993. *White Women, Race Matters: The Social Construction of Whiteness*. Minneapolis: University of Minnesota Press.

Galemba, Rebecca. 2012. "'Corn Is Food, Not Contraband': The Right to 'Free Trade' at the Mexico-Guatemala Border." *American Ethnologist* 39 (4): 716–34. http://dx.doi.org/10.1111/j.1548-1425.2012.01391.x

Galemba, Rebecca. 2013. "Illegality and Invisibility at Margins and Borders." *Political and Legal Anthropology Review* 36 (2): 274–85. http://dx.doi.org/10.1111/plar.12027

Gibbins, Roger. 1989. *Canada as a Borderlands Society*. Orono, ME: The Canadian-American Center.

Gilbert, Emily. 2007. "Leaky Borders and Solid Citizens: Governing Security, Prosperity and Quality of Life in a North American Partnership." *Antipode* 39 (1): 77–98. http://dx.doi.org/10.1111/j.1467-8330.2007.00507.x

Gilbert, Emily. 2012. "Borders and Security in North America." In *North America in Question: Regional Integration in an Era of Economic Turbulence*, edited by Jeffrey Ayres and Laura MacDonald, 196–217. Toronto: University of Toronto Press.

Goldring, Luin, Carolina Berinstein, and Judith K. Bernhard. 2009. "Institutionalizing Precarious Migratory Status in Canada." *Citizenship Studies* 13 (3): 239–65. http://dx.doi.org/10.1080/13621020902850643

Granatstein, Jack L. K. 1996. *Yankee Go Home? Canadians and Anti-Americanism*. Toronto: HarperCollins.

Graybill, Andrew R., and Benjamin H. Johnson. 2010. *Bridging National Borders in North America*. Durham, NC: Duke University Press.

Green, Sarah. 2012. "A Sense of Border." In *A Companion to Border Studies*, edited by Thomas Wilson and Hastings Donnan, 573–92. Malden, MA: Wiley-Blackwell. http://dx.doi.org/10.1002/9781118255223.ch33

Greenberg, Jessica. 2011. "On the Road to Normal: Negotiating Agency and State Sovereignty in Postsocialist Serbia." *American Anthropologist* 113 (1): 88–100. http://dx.doi.org/10.1111/j.1548-1433.2010.01308.x

Grinde, Donald A. (Yamasee). 2002. "Iroquois Border Crossings: Place, Politics, and the Jay Treaty." In *Globalization in the Line: Culture, Capital, and Citizenship at U.S. Borders*, edited by Claudia Sadowski-Smith, 167–80. New York: Palgrave.

Hale, Geoffrey. 2011. "Politics, People and Passports: Contesting Security, Travel and Trade on the US-Canadian Border." *Geopolitics* 16 (1): 27–69. http://dx.doi.org/10.1080/14650045.2010.493768

Hardwick, Susan W. 2010. "Fuzzy Transnationals? American Settlement, Identity, and Belonging in Canada." *American Review of Canadian Studies* 40 (1): 86–103. http://dx.doi.org/10.1080/02722010903536953

Hardwick, Susan W., and Ginger Mansfield. 2009. "Discourse, Identity, and 'Homeland as Other' at the Borderlands." *Annals of the Association of American Geographers* 99 (2): 383–405. http://dx.doi.org/10.1080/00045600802708408

Hardwick, Susan W., and Heather Smith. 2012. "Crossing the 49th Parallel: American Immigrants in Canada and Canadians in the US." In *Immigrant Geographies of North American Cities*, edited by Carlos Teixeira, Wei Li, and Audrey Kobayashi, 288–311. Don Mills, ON: Oxford University Press.

Harvard Law School. 2006. "Bordering on Failure: The U.S.-Canada Safe Third Country Agreement Fifteen Months after Implementation." Report prepared for the Harvard Immigration and Refugee Law Clinical Program. Cambridge, MA: Harvard Law School. Accessed 26 March 2014, http://hrp.law.harvard.edu/wp-content/uploads/2013/11/Harvard_STCA_Report.pdf

Heininen, Lassi, and Michele Zebich-Knos. 2011. "Comparing Arctic and Antarctic Border Debates." In *The Ashgate Research Companion to Border Studies*, edited by Doris Wastl, 195–217. Farnham, UK: Ashgate.

Hele, Karl, ed. 2008. *Lines Drawn Upon the Water: The First Nations Experience in the Great Lakes Borderlands*. Waterloo, ON: Wilfrid Laurier University Press.

Helleiner, Eric. 2006. *Towards North American Monetary Union? The Politics and History of Canada's Exchange Rate Regime*. Montreal: McGill-Queen's University Press.

Henry, Frances, and Carol Tator. 2010. *The Colour of Democracy: Racism in Canadian Society*. Toronto: Nelson.

Hess, Julia Meredith, and Dianna Shandy. 2008. "Kids at the Crossroads: Global Childhood and the State." *Anthropological Quarterly* 81 (4): 765–776. http://dx.doi.org/10.1353/anq.0.0035

Heyman, Josiah McC. 2008. "Constructing a Virtual Wall: Race and Citizenship in U.S.-Mexico Border Policing." *Journal of the Southwest* 50 (3): 305–34.

Heyman, Josiah McC. 2010 "US-Mexico Border Culture and the Challenge of Asymmetrical Interpenetration." In *Borderlands: Ethnographic Approaches to Security, Power and Identity*, edited by Hastings Donnan and Thomas Wilson, 21–34. Lanham, MD: University Press of America.

Heyman, Josiah McC. 2012. "Culture Theory and the US-Mexico Border." In *A Companion to Border Studies*, edited by Thomas Wilson and Hastings Donnan, 48–65. Malden, MA: Wiley-Blackwell.

Hickey, Robert. 2008. *Challenges and Opportunities for Improving Employment Conditions in Niagara's Hotel Sector*. Toronto: Unite Here! Canada.

Hier, Sean P., and Joshua L. Greenberg. 2002. "Constructing a Discursive Crisis: Crisis Problematization and *Illegal* Chinese in Canada." *Ethnic and Racial Studies* 25 (3): 490–513. http://dx.doi.org/10.1080/01419870020036701

Hipfl, Brigitte, Anita Bister, and Petra Strohmaier. 2003. "Youth Identities along the Eastern Border of the European Union." *Journal of Ethnic and Migration Studies* 29 (5): 835–48. http://dx.doi.org/10.1080/1369183032000149596

International Civil Liberties Monitoring Group. 2010. *Report of the Information Clearinghouse on Border Controls and Infringements to Travellers' Rights.*

Accessed 26 March 2014, http://www.travelwatchlist.ca/updir/travelwatchlist/ICLMG_Watchlists_Report.pdf

Jackson, John. 2003. *The Mighty Niagara: One River-Two Frontiers*. Amherst, NY: Prometheus Books.

Jameson, Elizabeth, and Jeremy Mouat. 2006. "Telling Differences: The Forty-Ninth Parallel and Historiographies of the West and Nation." *Pacific Historical Review* 75 (2): 183–230. http://dx.doi.org/10.1525/phr.2006.75.2.183

Jamil, Uzma, and Cecile Rousseau. 2012. "Subject Positioning, Fear, and Insecurity in South Asian Muslim Communities in the War on Terror Context." *Canadian Review of Sociology* 49 (4): 370–88. http://dx.doi.org/10.1111/j.1755-618X.2012.01299.x

Jefferess, David. 2009. "Responsibility, Nostalgia, and the Mythology of Canada as a Peacekeeper." *University of Toronto Quarterly* 78 (2): 709–27. http://dx.doi.org/10.3138/utq.78.2.709

Jiwani, Yasmin. 2009. "Helpless Maidens and Chivalrous Knights: Afghan Women in the Canadian Press." *University of Toronto Quarterly* 78 (2): 728–44. http://dx.doi.org/10.3138/utq.78.2.728

Johnson, Corey, Reece Jones, Anssi Paasi, Louise Amoore, Alison Mountz, Mark Salter, and Chris Rumford. 2011. "Interventions on Rethinking 'The Border' in Border Studies." *Political Geography* 30 (2): 61–9. http://dx.doi.org/10.1016/j.polgeo.2011.01.002

Jones, Angela. 2011. *African American Civil Rights: Early Activism and the Niagara Movement*. Santa Barbara, CA: Praeger.

Kazimi, Ali. 2012. *Undesirables: White Canada and the Komagata Maru: An Illustrated History*. Vancouver: Douglas and McIntyre.

Kearney, Michael. 1995. "The Local and the Global: The Anthropology of Globalization and Transnationalism." *Annual Review of Anthropology* 24 (1): 547–65. http://dx.doi.org/10.1146/annurev.an.24.100195.002555

Khosravi, Shahram. 2010. *"Illegal" Traveller: An Auto-Ethnography of Borders*. Basingstoke, UK: Palgrave MacMillan. http://dx.doi.org/10.1057/9780230281325

Kitossa, Tamari, and Katerina Deliovsky. 2010. "Interracial Unions with White Partners and Racial Profiling: Experiences and Perspectives." *International Journal of Criminology and Sociological Theory* 3 (2): 512–30.

Klug, Thomas. 2010. "The Immigration and Naturalization Service (INS) and the Making of a Border-Crossing Culture on the US-Canada Border, 1891–1941." *American Review of Canadian Studies* 40 (3): 395–415.http://dx.doi.org/10.1080/02722011.2010.496903

Kobayashi, Addie. 1998. *Exiles in Our Own Country: Japanese Canadians in Niagara*. Richmond Hill, ON: Nikkei Network of Niagara.

Konrad, Victor. 2012. "Conflating Imagination, Identity, and Affinity in the Social Construction of Borderlands Culture Between Canada and the United States." *American Review of Canadian Studies* 42 (4): 530–48. http://dx.doi.org/10.1080/02722011.2012.732096

Konrad, Victor, and Heather Nicol. 2008. *Beyond Walls: Reinventing the Canada/United States Borderlands*. Aldershot, UK: Ashgate.

Konrad, Victor, and Heather Nicol. 2011. "Border Culture, the Boundary between Canada and the United States of America, and the Advancement of Borderlands Theory." *Geopolitics* 16 (1): 70–90. http://dx.doi.org/10.1080/14650045.2010.493773

Kymlicka, Will. 2003. "Being Canadian." *Government and Opposition* 38 (3): 357–85. http://dx.doi.org/10.1111/1477-7053.t01-1-00019

Lask, Thomas. 1994. "'Baguettte-Heads' and 'Spiked Helmets': Children's Constructions of Nationality at the German-French Border." In *Border Approaches: Anthropological Perspectives on Frontiers*, edited by Hastings Donnan and Tom M. Wilson, 63–73. Landham, MD: University Press of America.

Laxer, James. 2003. *The Border: Canada, the U.S. and Dispatches from the 49th Parallel*. Toronto: Doubleday Canada.

Lugo, Alejandro. 2000. "Theorizing Border Inspections." *Cultural Dynamics* 12 (3): 353–73. http://dx.doi.org/10.1177/092137400001200305

Liebmann, Coleen. 2002. *Violations of Migrants Rights at the Border*. San Francisco, CA: International Human Rights Clinic.

Luibhéid, Eithne. 2002. *Entry Denied: Controlling Sexuality at the Border*. Minneapolis: University of Minnesota Press.

Lyon, David, and David Murakami Wood. 2012. "Security, Surveillance, and Sociological Analysis." *Canadian Review of Sociology* 49 (4): 317–27. http://dx.doi.org/10.1111/j.1755-618X.2012.01297.x

Lyon, David. 2003. *Surveillance as Social Sorting: Privacy, Risk and Digital Discrimination*. New York: Routledge.

Lwebuga-Mukasa, Jamson, S. Tonny Oyana, Arun Thenappan, and Sanjay J. Ayirookuzhi. 2004. "Association between Traffic Volume and Health Care Use for Asthma among Residents at a U.S-Canadian Border Crossing Point." *Journal of Asthma* 41 (3): 289–304. http://dx.doi.org/10.1081/JAS-120026086

Mackey, Eva. 1999. *The House of Difference: Cultural Politics and National Identity in Canada*. London: Routledge.

Macpherson, Alan D., and James McConnell. 2007. "A Survey of Cross-Border Trade at a Time of Heightened Security: The Case of the Niagara Bi-National Region." *American Review of Canadian Studies* 37 (3): 301–21. http://dx.doi.org/10.1080/02722010709481805

Mannik, Lynda. 2013. *Photography, Memory, and Refugee Identity: the Voyage of the S.S. Walnut, 1948*. Vancouver: University of British Columbia Press.

Mannik, Lynda. 2014. "Remembering Arrivals of Refugees by Boat in a Canadian Context." *Memory Studies* 7 (1): 76–91. http://dx.doi.org/10.1177/1750698013501361

Manore, Jean. 2011. "The Historical Erasure of an Indigenous Identity in the Borderlands: The Western Abenaki of Vermont, New Hampshire and Quebec." *Journal of Borderland Studies* 26 (2): 179–96. http://dx.doi.org/10.1080/08865655.2011.641320

McDougall, Allan K., and Lisa Philips. 2012. "The State, Hegemony and the Historical British-US Border." In *A Companion to Border Studies*, edited by Thomas Wilson and Hastings Donnan, 177–93. Malden, MA: Wiley-Blackwell. http://dx.doi.org/10.1002/9781118255223.ch10

McGreevy, Patrick V. 1988. "The End of America: The Beginning of Canada." *Canadian Geographer* 32 (4): 307–18.

McGreevy, Patrick V. 1991. *The Wall of Mirrors: Nationalism and Perceptions of the Border at Niagara Falls*, 1–18. Orono, ME: Borderlands.

McGreevy, Patrick V. 1994. *Imagining Niagara: The Meaning and Making of Niagara Falls*. Amherst: University of Massachusetts Press.

McKittrick, Katherine. 2007. "Freedom is a Secret." In *Black Geographies and the Politics of Place*, edited by Katherine McKittrick and Clyde Woods, 97–114. Cambridge, MA: South End Press.

McManus, Sheila. 2005. *The Line Which Separates: Race, Gender and the Making of the Alberta-Montana Borderlands*. Edmonton: University of Alberta Press.

Meyers, Deborah Waller, and Demetrios G. Papademetriou. 2001. "Self-Governance along the U.S.-Canada Border: A View from Three Regions." In *Caught in the Middle: Border Communities in an Era of Globalization*, edited by Demetrios G. Papademetriou and Deborah Waller Meyers, 44–87. Washington: Carnegie Endowment for International Peace.

Mountz, Alison. 2010. *Seeking Asylum: Human Smuggling and Bureaucracy at the Border*. Minneapolis: University of Minnesota Press. http://dx.doi.org/10.5749/minnesota/9780816665372.001.0001

Mountz, Alison. 2011. "Specters at the Port of Entry: Understanding State Mobilities through an Ontology of Exclusion." *Mobilities* 6 (3): 317–34. http://dx.doi.org/10.1080/17450101.2011.590033

Mountz, Alison, and Nancy Hiemstra. 2012. "Spatial Strategies for Rebordering Human Migration at Sea." In *A Companion to Border Studies*, edited by Thomas Wilson and Hastings Donnan, 455–72. Malden, MA: Wiley-Blackwell. http://dx.doi.org/10.1002/9781118255223.ch26

Munck, R. 2008. "Globalisation, Governance and Migration: An Introduction." *Third World Quarterly* 29 (7): 1227–46. http://dx.doi.org/10.1080/01436590802386252

Murray, David. 2002. *Colonial Justice: Justice, Morality, and Crime in the Niagara District, 1791–1849*. Toronto: University of Toronto Press.

Neumayer, Eric. 2006. "Unequal Access to Foreign Spaces: How States Use Visa Restrictions to Regulate Mobility in a Globalized World." *Transactions of the Institute of British Geographers* 31 (1): 72–84. http://dx.doi.org/10.1111/j.1475-5661.2006.00194.x

Neumayer, Eric. 2010. "Visa Restrictions and Bilateral Travel." *Professional Geographer* 62 (2): 171–81. http://dx.doi.org/10.1080/00330121003600835

Nevins, Joseph. 2007. "Dying for a Cup of Coffee: Migrant Deaths in the U.S.-Mexico Border Region in a Neoliberal Age." *Geopolitics* 12 (2): 228–47. http://dx.doi.org/10.1080/14650040601168826

Newman, David. 2011. "Contemporary Research Agendas in Border Studies: An Overview." In *The Ashgate Research Companion to Border Studies*, edited by Doris Wastl-Walter, 33–47. Farnham, UK: Ashgate.

Ngai, Mae. 2004. *Impossible Subjects: Illegal Aliens and the Making of Modern America*. Princeton, NJ: Princeton University Press.

Nicks, Joan, and Barry Keith Grant, eds. 2010. *Covering Niagara: Studies in Local Popular Culture*. Waterloo, ON: Wilfrid Laurier University Press.

Nicol, Heather. 2005. "Resiliency or Change? The Contemporary Canada-US Border." *Geopolitics* 10 (4): 767–90. http://dx.doi.org/10.1080/14650040500318597

Nicol, Heather. 2011. "Building Borders the Hard Way: Enforcing North American Security Post 9/11." In *The Ashgate Research Companion to Border Studies*, edited by Doris Wastl-Walter, 263–82. Farnham, UK: Ashgate.

Newman, David. 2006. "The Lines That Continue to Separate Us: Borders in Our 'Borderless' World." *Progress in Human Geography* 30 (2): 143–61.

Nyers, Peter. 2010. "Abject Cosmopolitanism: The Politics of Protection in the Anti-deportation Movement." In *The Deportation Regime: Sovereignty, Space and the Freedom of Movement*, edited by Nicholas de Genova and Natalie Peutz, 413–41. Durham, NC: Duke University Press. http://dx.doi.org/10.1215/9780822391340-016

O'Connell, Anne. 2010. "An Exploration of Redneck Whiteness in Multicultural Canada." *Social Politics* 17 (4): 536–63. http://dx.doi.org/10.1093/sp/jxq019

Paasi, Anssi. 2011. "A Border Theory: An Unattainable Dream or a Realistic Aim for Border Scholars?" In *The Ashgate Research Companion to Border Studies*, edited by Doris Wastl-Walter, 11–31. Farnham, UK: Ashgate.

Park, Hijin. 2010. "A Nation through Home Invasion: White Settler Nationalism, Chinese Criminality and 1999 'Summer of the Boats.'" *Asian Journal of Canadian Studies* 16 (2): 89–113.

Paulus, Jeremy, and Ali Asgary. 2010. "Enhancing Border Security: Local Values and Preferences at the Blue Water Bridge (Point Edward, Canada)." *Journal of Homeland Security and Emergency Management* 7 (1): 1–26. http://dx.doi.org/10.2202/1547-7355.1543

Pew Research Center. 2005. *U.S. Image Up Slightly, But Still Negative.* Washington: Pew Research Center. Accessed 26 March 2014, http://www.pewglobal.org/2005/06/23/us-image-up-slightly-but-still-negative

Pickering, Sharon, and Julie Ham. 2014. "Hot Pants at the Border: Sorting Sex Worker from Trafficking." *British Journal of Criminology* 54 (1): 2–19. http://dx.doi.org/10.1093/bjc/azt060

Popescu, Gabriel. 2012. *Bordering and Ordering the Twenty First Century: Understanding Borders.* Lanham, MD: Rowman and Littlefield.

Pottie-Sherman, Yolande, and Rima Wilkes. 2015. "Visual Media and the Construction of the Benign Canadian Border on National Geographic's *Border Security.*" *Social and Cultural Geography* (3) 1: 1–20. http://dx.doi.org/10.1080/14649365.2015.1042400

Pratt, Anna. 2005. *Securing Borders.* Vancouver: University of British Columbia Press.

Pratt, Anna. 2010. "Between a Hunch and a Hard Place: Making Suspicion Reasonable at the Canadian Border." *Social and Legal Studies* 19 (4): 461–80. http://dx.doi.org/10.1177/0964663910378434

Pratt, Anna, and Sara K. Thompson. 2008. "'Chivalry, 'Race' and Discretion at the Canadian Border." *British Journal of Criminology* 48 (5): 620–40. http://dx.doi.org/10.1093/bjc/azn048

Price, John. 2011. *Orienting Canada: Race, Empire and the Transpacific.* Vancouver: University of British Columbia Press.

Price, John. 2013. "Canada, White Supremacy, and the Twinning of Empires." *International Journal* 68 (4): 628–38. http://dx.doi.org/10.1177/0020702013510675

Ramirez, Bruno. 2001. *Crossing the 49th Parallel: Migration from Canada to the United States, 1900–1930.* Ithaca, NY: Cornell University Press.

Razack, Sherene. 2002. "When Race Becomes Place." In *Race, Space, and the Law: Unmapping a White Settler Society*, edited by Sherene Razack, 1–20. Toronto: Between the Lines.

Razack, Sherene. 2004. *Dark Threats and White Knights: The Somalia Affair, Peacekeeping, and the New Imperialism.* Toronto: University of Toronto Press.

Razack, Sherene. 2009. "Afterword: Race, Desire, and Contemporary Security Discourses." *University of Toronto Quarterly* 78 (2): 815–20. http://dx.doi.org/10.3138/utq.78.2.815

Razack, Sherene. 2010. "Abandonment and the Dance of Race and Bureaucracy in Spaces of Exception." In *States of Race: Critical Race Feminism for the Twenty-First Century*, edited by Sherene Razack, Malinda Smith, and Sunera Thobani, 87–107. Toronto: Between the Lines.

Razack, Sherene, Malinda Smith, and Sunera Thobani, eds. 2010. *States of Race: Critical Race Feminism for the Twenty-First Century*. Toronto: Between the Lines.

Regional Institute. 2007. *Defining the Region's Edge*. Buffalo: University at Buffalo Regional Institute. Accessed 20 December 2015, http://regional-institute.buffalo.edu/wp-content/uploads/sites/3/2014/10/Defining-the-Regions-Edge-Policy-Brief.pdf

Revie, Linda Lee. 2003. *The Niagara Companion: Explorers, Artists and Writers at the Falls, from Discovery through the Twentieth Century*. Waterloo, ON: Wilfrid Laurier University Press.

Rippberger, Susan, and Kathleen Staudt. 2003. *Pledging Allegiance: Learning Nationalism at the El Paso-Juarez Border*. New York: Routledge Farmer.

Roberts, Gillian, and David Stirrup, eds. 2013. *Parallel Encounters: Culture at the Canada-US Border*. Waterloo, ON: Wilfrid Laurier University Press.

Rosaldo, Renato. 1993. *Culture and Truth: The Remaking of Social Analysis*. Boston: Beacon Press.

Sadowski-Smith, Claudia. 2014. "The Centrality of the Canada-US Border for Hemispheric Studies of the Americas." *Forum for Inter-American Research* 7 (3): 20–40.

Salter, Mark B. 2006. "The Global Visa Regime and the Political Technologies of the International Self: Borders, Bodies, Biopolitics." *Alternatives: Local, Global, Political* 31 (2): 167–89. http://dx.doi.org/10.1177/030437540603100203

Salter, Mark. 2007. "Canadian Post-9/11 Border Policy and Spillover Securitization: Smart, Safe, and Sovereign?" In *Critical Policy Studies*, edited by Michael Orsini and Miriam Smith, 299–319. Vancouver: University of British Columbia Press.

Salter, Mark B., and Genevieve Piche. 2011. "The Securitization of the US-Canada Border in American Political Discourse." *Canadian Journal of Political Science* 44 (4): 929–51. http://dx.doi.org/10.1017/S0008423911000813

Satzewich, Vic. 2015. *Points of Entry: How Canada's Immigration Officers Decide Who Gets In*. Vancouver: University of British Columbia Press.

Scheper-Hughes, Nancy, and Carolyn Sargent, eds. 1998. *Small Wars: The Cultural Politics of Childhood*. Berkeley: University of California Press.

Schneekloth, Lynda, and Robert Shibley. 2005. "Imagine Niagara." *Journal of Canadian Studies. Revue d'Etudes Canadiennes* 39 (3): 105–20.

Sharma, Nandita. 2006a. "White Nationalism, Illegality and Imperialism: Border Controls as Ideology." In *(En)Gendering the War on Terror: War Stories and Camouflaged Politics*, edited by Krista Hunt and Kim Rygiel, 121–143. Aldershot, UK: Ashgate.

Sharma, Nandita. 2006b. *Home Economics: Nationalism and the Making of "Migrant Workers" in Canada*. Toronto: University of Toronto Press.

Shahzad, Farhad. 2012. "Forging the Nation as an Imagined Community." *Nations and Nationalism* 18 (1): 21–38. http://dx.doi.org/10.1111/j.1469-8129.2011.00502.x

Sheller, Mimi, and John Urry. 2006. "The New Mobilities Paradigm." *Environment and Planning A* 38 (2): 207–26. http://dx.doi.org/10.1068/a37268

Shepard, R. Bruce. 1991. "Plain Racism: The Reaction against Oklahoma Black Immigrants to the Canadian Plains." In *Racism in Canada*, edited by Ormond McKague, 15–31. Saskatoon: Fifth House.

Siener, William. 2008a. "Through the Back Door: Evading the Chinese Exclusion Act along the Niagara Frontier, 1900–1924." *Journal of American Ethnic History* 27 (4): 34–70.

Siener, William. 2008b. "'A Barricade of Ships, Guns, Airplanes and Men': Arming the Niagara Border, 1920–1930." *American Review of Canadian Studies* 38 (4): 429–50.

Siener, William. 2013. "The United Nations at Niagara: Borderlands Collaboration and Emerging Globalism." *American Review of Canadian Studies* 43 (3): 377–93. http://dx.doi.org/10.1080/02722011.2013.819583

Simpson, Audra. 2008. "Subjects of Sovereignty: Indigeneity, the Revenue Rule, and Juridics of Failed Consent." *Law and Contemporary Problems* 71 (3): 191–215.

Simpson, Audra. 2014. *Mohawk Interruptus: Political Life across the Borders of Settler States*. Durham, NC: Duke University Press. http://dx.doi.org/10.1215/9780822376781

Singleton, Sara. 2009. *Not Our Borders: Indigenous People and the Struggle to Maintain Shared Lives and Cultures in Post-9/11 North America*. Working Paper No. 4. Bellingham: Border Policy Research Institute, Western Washington University.

Slack, Jeremy, and Scott Whiteford. 2011. "Violence and Migration on the Arizona-Sonora Border." *Human Organization* 70 (1): 11–21. http://dx.doi.org/10.17730/humo.70.1.k34n00130470113w

Sohi, Seema. 2011. "Race, Surveillance, and Indian Anticolonialism in the Transnational Western U.S.-Canadian Borderlands." *Journal of American History* 98 (2): 420–36. http://dx.doi.org/10.1093/jahist/jar210

Sparke, Matthew. 2006. "A Neoliberal Nexus: Economy, Security and the Biopolitics of Citizenship on the Border." *Political Geography* 25 (2): 151–80. http://dx.doi.org/10.1016/j.polgeo.2005.10.002

Spener, D. 2009. *Clandestine Crossings: Migrants and Coyotes on the Texas-Mexico Border*. Ithaca, NY: Cornell University Press.

Spyrou, Spyros, and Miranda Christou, eds. 2014. *Children and Borders. Houndsmills*. Basingstoke, UK: Palgrave MacMillan.

Stephens, Sharon, ed. 1995. *Children and the Politics of Culture*. Princeton, NJ: Princeton University Press.

Stieve, Thomas. 2005. "National Identity in Niagara Falls, Canada and the United States." *Tijdschrift voor Economische en Sociale Geografie* 96 (1): 3–14. http://dx.doi.org/10.1111/j.1467-9663.2005.00435.x

Stronza, Amanda. 2001. "Anthropology of Tourism: Forging New Ground for Ecotourism and Other Alternatives." *Annual Review of Anthropology* 30 (1): 261–83. http://dx.doi.org/10.1146/annurev.anthro.30.1.261

Tanovich, David. 2006. *The Colour of Justice: Policing Race in Canada*. Toronto: Irwin Law.

Tator, Carol, and Frances Henry. 2006. *Racial Profiling in Canada: Challenging the Myth of "A Few Bad Apples."* Toronto: University of Toronto Press.

Thobani, Sunera. 2007. *Exalted Subjects*. Toronto: University of Toronto Press.

Traister, Bryce. 2002. "Border Shopping, American Studies and the Anti-Nation." In *Globalization on the Line: Culture, Capital and Citizenship at U.S. Borders*, edited by Claudia Sadowski-Smith, 31–52. New York: Palgrave.

Turbeville, Daniel E., and Susan L. Bradbury. 2005. "NAFTA and Transportation Corridor Improvement in Western North America: Restructuring for the Twenty-First Century." In *Holding the Line: Borders in a Global World*, edited by Heather Nicol and Ian Townsend-Gault, 268–90. Vancouver: University of British Columbia Press.

Truitt, Allison. 2008. "On the Back of a Motorbike: Middle-Class Mobility in Ho Chi Minh City." *American Ethnologist* 35 (1): 3–19. http://dx.doi.org/10.1111/j.1548-1425.2008.00002.x

United States Government Accountability Office. 2010. "Border Security. Enhanced DHS Oversight and Assessment of Interagency Coordination is Needed for Northern Border." Report to Congressional Requesters. Accessed 1 August 2011, http://www.gao.gov/new.items/d1197.pdf

Van Houtum, Henk. 2010. "Human Blacklisting: The Global Apartheid of the EU's External Border Regime." *Environment and Planning. D, Society and Space* 28 (6): 957–76. http://dx.doi.org/10.1068/d1909

Van Liempt, Ilse, and Stephanie Sersli. 2013. "State Responses and Migrant Experiences with Human Smuggling: A Reality Check." *Antipode* 45 (4): 1029–46. http://dx.doi.org/10.1111/j.1467-8330.2012.01027.x

Vila, Pablo. 2000. *Crossing Borders, Reinforcing Borders*. Austin: University of Texas Press.

Vila, Pablo, ed. 2003. *Ethnography at the Border*. Minneapolis: University of Minnesota Press.

Vila, Pablo. 2005. *Border Identifications*. Austin: University of Texas Press.

Yuval-Davis, Nira, and Marcel Stoetzler. 2002. "Imagined Boundaries and Borders: A Gendered Gaze." *European Journal of Women's Studies* 9 (3): 329–44. http://dx.doi.org/10.1177/1350506802009003378

Walcott, Rinaldo. 2003. *Black Like Who? Writing Black Canada*. 2nd rev. ed. Toronto: Insomniac Press.

Wastl-Walter, Doris, ed. 2011. *The Ashgate Research Companion to Border Studies*. Farnham, UK: Ashgate.

Whitaker, Reg, Gregory S. Kealey, and Andrew Parnaby. 2012. *Secret Service: Political Policing in Canada: From the Fenians to Fortress America*. Toronto: University of Toronto Press.

Widdis, Randy. 2010. "Crossing an Intellectual and Geographic Border: The Importance of Migration in Shaping Canadian-American Borderlands at the Turn of the Twentieth Century." *Social Science History* 34 (4): 445–97. http://dx.doi.org/10.1215/01455532-2010-012

Wilkes, Rima, Catherine Corrigall-Brown, and Danielle Ricard. 2010. "Nationalism and Media Coverage of Indigenous People's Collective Action in Canada." *American Indian Culture and Research Journal* 34 (4): 41–59. http://dx.doi.org/10.17953/aicr.34.4.05q6170n1m987792

Wigmore, Gregory. 2011. "Before the Railroad: From Slavery to Freedom in the Canadian-American Borderland." *Journal of American History* 98 (2): 437–54. http://dx.doi.org/10.1093/jahist/jar256

Williamson, Ronald, and Robert MacDonald. 1998. *Legacy of Stone: Ancient Life on the Niagara Frontier*. Toronto: Eastendbooks.

Wilson, Thomas, and Hastings Donnan, eds. 1998. *Border Identities: Nation and State as International Frontiers*. Cambridge: Cambridge University Press. http://dx.doi.org/10.1017/CBO9780511607813

Wilson, Thomas, and Hastings Donnan. 2012a. "Borders and Border Studies." In *A Companion to Border Studies*, edited by Thomas Wilson and Hastings Donnan, 1–25. Malden, MA: Wiley-Blackwell. http://dx.doi.org/10.1002/9781118255223.ch1

Wilson, Thomas, and Hastings Donnan, eds. 2012b. *A Companion to Border Studies*. Malden, MA: Wiley-Blackwell. http://dx.doi.org/10.1002/9781118255223

Wimmer, Andreas, and Nina Glick Schiller. 2002. "Methodological Nationalism and Beyond: Nation-State Building, Migration and the Social Sciences." *Global Networks* 2 (4): 301–34. http://dx.doi.org/10.1111/1471-0374.00043

Winant, Howard. 2004. *The New Politics of Race: Globalism, Difference, Justice.* Minneapolis: University of Minnesota Press.

Winter, Elke. 2007. "Neither 'America' nor 'Québec': Constructing the Canadian Multicultural Nation." *Nations and Nationalism* 13 (3): 481–503. http://dx.doi.org/10.1111/j.1469-8129.2007.00295.x

Winter, Elke. 2014. "(Im)possible Citizens: Canada's 'Citizenship Bonanza' and its Boundaries." *Citizenship Studies* 18 (1): 46–62. http://dx.doi.org/10.1080/13621025.2012.707010

Winter, Elke. 2015. "Rethinking Multiculturalism after Its 'Retreat': Lessons from Canada." *American Behavioral Scientist* 59 (6): 637–57. http://dx.doi.org/10.1177/0002764214566495

Wortley, Scott, and Julian Tanner. 2005. "Inflammatory Rhetoric? Baseless Accusations? A Response to Gabor's Critique of Racial Profiling Research in Canada." *Canadian Journal of Criminology and Criminal Justice* 47 (3): 581–610. http://dx.doi.org/10.3138/cjccj.47.3.581

Zelizer, Viviana. 1998. "Introduction: How People Talk about Money." *American Behavioral Scientist* 41 (10): 1373–83. http://dx.doi.org/10.1177/0002764298041010002

Ziolkowski, Michael. 2006. "Perceptions of Border Security along the Canada-U.S. Border, Grand Island, New York." Canada-United States Trade Center Occasional Paper No. 34. Buffalo, NY: Canada-United States Trade Center.

Index

www.ingramcontent.com/pod-product-compliance
Lightning Source LLC
LaVergne TN
LVHW090158080826
844660LV00013B/853/J

* 9 7 8 1 4 4 2 6 4 9 0 5 7 *